W9-ARA-182

The Eye of the Elephant

Books by Delia and Mark Owens

CRY OF THE KALAHARI

THE EYE OF THE ELEPHANT
An Epic Adventure in the African Wilderness

SECRETS OF THE SAVANNA
Twenty-three Years in the African Wilderness
Unraveling the Mysteries of Elephants and People

The Eye of the
ELEPHANT

An Epic Adventure in the
African Wilderness

DELIA AND MARK OWENS

A MARINER BOOK
HOUGHTON MIFFLIN COMPANY
BOSTON · NEW YORK

Copyright © 1992 by Delia and Mark Owens
All rights reserved

For information about permission to reproduce selections from
this book, write to Permissions, Houghton Mifflin Company,
215 Park Avenue South, New York, New York 10003.

Library of Congress Cataloging-in-Publication Data
Owens, Delia.
  The eye of the elephant : an epic adventure in the African
wilderness / Delia and Mark Owens.
  p. cm.
Includes bibliographical references and index.
ISBN-13: 978-0-395-42381-3   ISBN-13: 978-0-395-68090-2 (pbk.)
ISBN-10: 0-395-42381-3   ISBN-10: 0-395-68090-5 (pbk.)
  1. Wildlife conservation—Luangwa River Valley (Zambia and
Mozambique) 2. Elephants—Luangwa River Valley (Zambia
and Mozambique) 3. Owens, Delia. 4. Owens, Mark.
  5. Wildlife conservationists—United States—Biography.
  I. Owens, Mark. II. Title
QL84.6.233084 1992
639.9'7961—dc20 92-17691 CIP

Printed in the United States of America

Maps by George Ward
Design by Melodie Wertelet

QUM 15 14 13 12 11 10 9 8

Permissions for quotations that appear
in this book are listed on page 306.

To Helen and Fred,

   Bobby and Mary,

   and Mama — for doing so much.

And to Lee and Glenda, who keep us all smiling.

o o o

Contents

Authors' Note

This story is not meant to judge Zambia's past conservation practices so much as to project hope for the future. The events described in this book occurred under the previous one-party Marxist government in Zambia. In 1991 the Zambian people elected a truly democratic government, which has taken positive steps to address the conservation problems of the country. It is only because of this change in government that we have the freedom to tell our story. Scientists and conservationists have the responsibility and the right to report their findings. By telling the truth, no matter how controversial, they incur a measure of personal and professional risk; by not telling it, we *all* risk much, much more.

The names of the innocent in this book have been changed to protect them from the guilty; the names of the guilty have been changed to protect us. The rest of this story is true.

Principal Characters

Island Zulu	the "Camp-in-Charge" at Mano Camp
Patrick Mubuka	the "Camp-in-Charge" at Nsansamina Camp
Nelson Mumba Gaston Phiri Tapa	the game scouts at Mano Camp
Chomba Simbeye Chanda Mwamba Mutale Kasokola	the first Bemba tribesmen to work for us
Mosi Salama Bornface Mulenga	game wardens at Mpika
Sunday Justice Mumanga Kasokola	our cooks
Jealous Mvula Bwalya Muchisa Musakanya Mumba	our earliest informants
Chanda Seven Bernard Mutondo Mpundu Katongo Chikilinti Simu Chimba	the most notorious poachers of North Luangwa National Park
John Musangu Kotela Mukendwa	unit leaders at Mano Camp
Banda Chungwe	the senior ranger at Mpika
Evans Mukuka	our first educational officer
Marie and Harvey Hill	friends in Mpika
Chief Mukungule	chief of the area west of the park
Chief Nabwalya	chief of the area between North and South Luangwa parks

Chief Chikwanda	chief of the area near Mpika
Max Saili	our community service officer
Tom and Wanda Canon	volunteers from Texas
Ian Spincer Edward North	assistants from the University of Reading

The Lions

Happy Sunrise Sage Stormy Saucy	the new members of the Blue Pride in the Kalahari
Serendipity Kora	lions of the Serendipity Pride along the Mwaleshi River
Bouncer	the male we radio collar on the plains

The Elephants

Survivor	the male elephant who comes into our camp
Cheers	the male elephant who sometimes accompanies Survivor
Camp Group	the group of male elephants that forages near our camp
One Tusk Misty Mandy Marula	a family unit of female elephants
Long Ear and her daughters	another family unit of female elephants

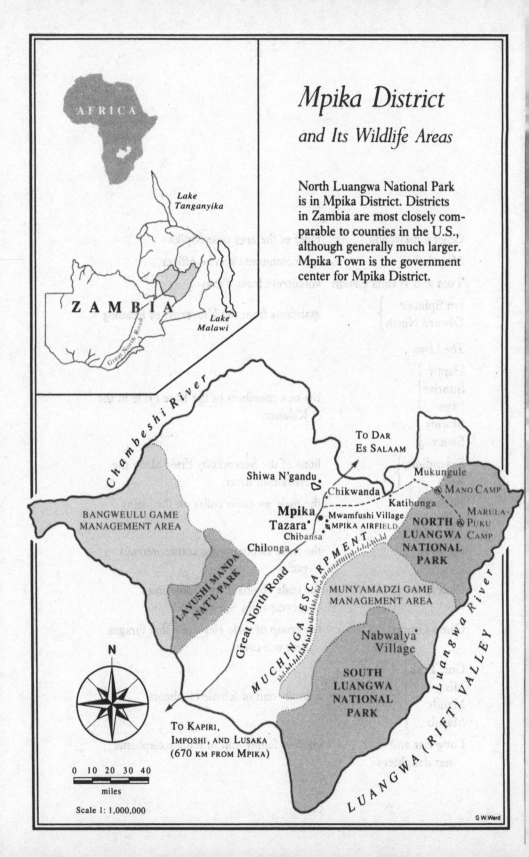

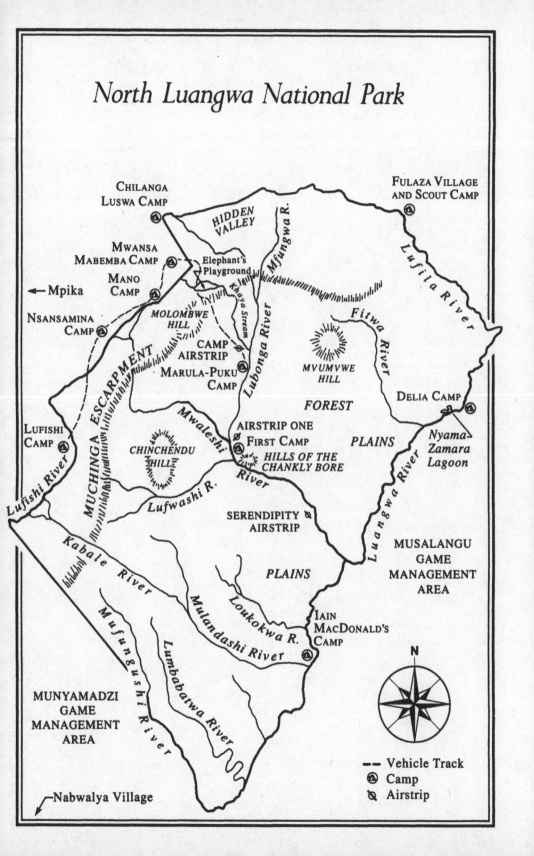

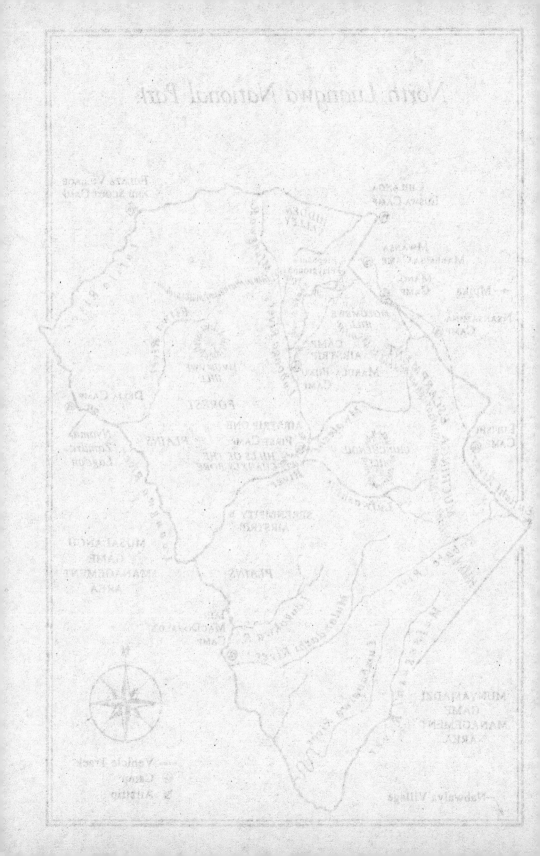

PART ONE ∘∘∘

The Dry Season

Prologue

DELIA

DAWN IN LUANGWA. I hear the elephant feeding on marula fruits just outside the cottage. Quietly pushing aside the mosquito net, I rise from the bed and tiptoe through the dark to the wash-room, which has a tiny window high under the thatched roof. All I can see in the window is a large eye, like that of a whale, blinking at me through the pale morning light.

One step at a time, I ease closer to the window until I am just below it. Then, standing on an old tea-crate cupboard, I pull myself up to the sill and see Survivor's eye only a foot away. Long, straight lashes partially cover his pupil as he looks toward the ground searching for a fruit. Then, as he picks one up with his trunk and puts it into his mouth, he lifts his lashes and looks directly at me. He shows neither surprise nor concern, and I stare into the gray forever of an elephant's eye.

Such an incident may take place in other areas of Africa, but not in the northern Luangwa Valley of Zambia. In the last fifteen years, one hundred thousand elephants have been slaughtered by poachers in this valley. Here elephants usually run at the first sight or scent of man. I want to remember always the deep furrows of folded skin above Survivor's lashes, his moist and glistening eye, which now reflects the sunrise. Surely this will never happen to me again; the memory must last a lifetime. And I must never forget the way I feel, for at this moment I can see everything so clearly.

o o o

We first came to Africa in 1974 and settled in Deception Valley, a dry, fossilized river in the Kalahari Desert of Botswana. For seven years we lived in tents among the bush-covered dunes, the only people other than a few scattered bands of Bushmen in a wilderness the size of Ireland. The lions and brown hyenas there apparently had never seen humans before. They accepted us into their prides and clans, revealing previously unknown details of their natural history. Our tree-island camp was in the center of the Blue Pride's territory. These lions — Blue, Sassy, Happy, Bones, and later Muffin and Moffet — often sat beyond our campfire or raided our pantry. Once, when sleeping on the open savanna, we awoke to find ourselves surrounded by lions an arm's length away.

We left Deception at the end of 1980 to complete our graduate work and returned in 1985, when this story begins. Our greatest hope was to find whichever Blue Pride lions might still be alive, and to continue the research for another five years. We would search every dune slope, dried water hole, and acacia grove until we found them.

But we had another objective, too. The Central Kalahari Game Reserve — long forgotten and ignored by the outside world — was now the center of controversy. Powerful cattlemen and politicians wanted to dissolve the reserve and divide it into large private ranches, even though the sandy desert savannas could not sustain cattle for long. We had a quite different recommendation: that the area be conserved for the benefit of the local people through wildlife tourism.

Despite the pressures on the Kalahari, surely few places on earth had changed so little during the four years we were away. There was still no development of any kind in the reserve. At our camp we would still have to haul water in drums for fifty miles, live in the same faded tents, drive on the one bush track we had made years before. Once again our only visitors would be lions, brown hyenas, jackals, springbok, giraffes, hornbills, and lizards.

Lost again among those dunes, we failed to realize that even though the Kalahari had remained much the same, the rest of Africa had changed. We had survived drought and sandstorms. Now we would be caught in another kind of storm — one that would uproot us and blow us like tumbleweeds across the continent in search of another wilderness. And there the storm would continue.

o o o

Survivor lowers his lashes again as he feels around for another fruit, finds one, and raises it to his mouth, smacking loudly as he chews. He looks back at me again. I can see not only into his eye, but through it. Beyond are thousands of elephants in massive herds wending their way along mountain trails and down into the valley, there to stroll slowly across stilled savannas surrounded by thick, luxuriant forests. Giant, gentle mothers and playful youngsters romp and bathe in wide, sweeping rivers, unafraid. Powerful males push and shove for courtship rights, then stand back from each other, shaking their heads, their ears flapping in a cloud of dust. Through Survivor's eye I can see the wilderness as it once was. The storm continues, but a ray of hope shines through. Because of it, some of wild Africa may be saved.

Slowly Survivor curls his trunk to the windowsill and takes in my scent as he looks directly at me again. I wiggle my fingers forward until they are pressed against the flyscreen, only inches from his trunk. I want to whisper something, but what could I say?

The eye of the elephant is the eye of the storm.

1

Flight to Deception

MARK

> Every time that I have gone up in an aeroplane and looking
> down have realized that I was free of the ground, I have
> had the consciousness of a great new discovery. "I see," I
> have thought. "This was the idea. And now I understand
> everything."
>
> — ISAK DINESEN

o o o

AIRBORNE OVER THE KALAHARI for the first time in years, I
felt as though I was meeting an old friend again in some secret
corner of the earth known only to the two of us. During our seven
years in this vast wilderness, I had got Kalahari sand in my shoes,
and civilization with its fine hotels, its restaurants, its hot baths
and other conveniences, had not been able to shake it out. The
farther north I flew, the farther into the desert. Seeing the familiar
pans, the fossil river valleys, the vast, undulating bush savannas
with giraffes browsing flat-topped *Acacia tortillas* trees, I knew I
was going back where I belonged. It was early April 1985.

While planning the flight to Botswana, I had been anxious that
the six-year drought might have so changed the Kalahari's features
that I would be unable to find our old camp. I was supposed to
have met Delia there two days ago, but last-minute problems with
the plane in Johannesburg had delayed me. During her long drive
into the Kalahari by truck, and even after she reached camp, there
was no way to alert her. If I didn't show up soon, she would think
I had been forced down somewhere.

Scanning the plane's instruments, my eyes locked onto the
gauge for my right tank. Halfway to Deception, its needle was
already nudging the red. I was losing fuel — fast. I straightened
up in my seat, looked along each wing for any sign of a leak, then

checked my carburetor mixture again. Nothing wrong there. Wiping my forearm across my eyes, I tried to stay calm.

My right tank was virtually dry while the left one read completely full, but I had set the fuel selector to draw equally from both. The line from the left tank to the engine must be blocked. If so, I would run out of gas within the next few minutes. I had to land immediately.

I looked out of the window and down 4500 feet. Six years of drought had flayed the Kalahari, the dry, hot winds searing all signs of life until the terrain looked like ground zero at the Nevada nuclear test flats: sterile, forbidding, unfamiliar. I swallowed hard, leaned forward in my seat, and began urgently looking for a place to make a precautionary landing. If I flew on, the engine might quit over bush savanna or woodlands, where a forced landing would end in an outright crash. No one would ever find me.

A perfectly round, brilliantly white salt pan appeared off to the left about fifteen miles away. I banked left and headed straight for it, pulling back the throttle to conserve avgas (aviation gasoline). The gauge for the right tank was now rock solid red, and several times the engine seemed to miss strokes. When the pan was finally below me, I took a breath and began setting up the plane for a landing.

But at 500 feet above the ground I noticed deep animal tracks in the surface. If I put the plane down here, its wheels would sink into the salt and powder. Even if I could find and fix the fuel problem, I would never get airborne again.

It occurred to me that I couldn't be positive that the left tank was blocked until the right one was dry. I would switch to the right tank and deliberately run it out of fuel while circling over the pan. If the engine quit I could land there safely, even if I was not able to take off again.

I circled overhead, waiting for the engine to die. It never did. The left tank began feeding fuel, its gauge slowly drawing down. Later I would learn that the plane's mechanics had cross-connected the lines from the fuel tanks to the fuel selector console. "Right" drew from the left tank; "Left" was drawing from the right. Worse, higher air pressure from a bad vent in the right tank was forcing its avgas into the left tank, bloating it. The excess was being pumped out through a leaky fuel cap on top of the left wing, where I couldn't see it. It took forty precious minutes of flying — and fuel — to figure all this out. Now even if I made a beeline for camp, I might not make it.

And my problems were just beginning. Within five minutes of leaving the pan, I realized I was lost. Nothing below me looked even vaguely familiar. Surely the drought could not have wiped out all my old landmarks. Where were the Khutse Pans, the "mitochondria" pans, the squiggles of fossil river that used to tell me my position in the desert? They were subtle, but four years ago I had known the Kalahari so well from the air that they were like road signs to me. Even though haze had cut my visibility to about two miles, it seemed impossible that I had flown past each of these features without seeing any of them. I tightened my grip on the controls and held my compass heading. Something familiar had to come along.

Forty-five minutes after leaving the pan, I was totally disoriented; and a stiff head wind had reduced my ground speed from 150 mph to 120. It would take even more fuel to get to camp. Desperate to see something — anything — recognizable, I spent precious avgas climbing to 9500 feet, where I hoped I could see farther over the desert. The result was the same. All below me was a whiteout of haze. I had to be miles off course, but which way I couldn't tell. The same mechanics who crossed the fuel lines had put a steel — rather than a nonmagnetic brass — screw in the compass housing. It was off by thirty degrees. But of course I didn't know it at the time.

I fought off the urge to leave my flight path to chase after smudges in the bleak landscape, hoping to find something familiar. I couldn't afford to gamble away the avgas. So I flew on, not daring to look at the gauges anymore.

An hour later I still had no idea where I was — and I knew for certain that I would run out of fuel before reaching camp. I could only hope that I would be near a Bushman village where I could get water, or at least some wild melons, to keep me alive. But I had seen none of the settlements that I knew from years ago. I must be many miles off course.

I switched the radio to 125.5, Botswana's civil air traffic frequency, and picked up the microphone. "Any aircraft listening on this frequency, this is Foxtrot Zulu Sierra. Do you read me?" There was no response. I repeated my sign several times, but the only answer was the hiss of static in my earphones.

I changed to 121.5, the emergency frequency, and called again: "Any aircraft, this is Foxtrot Zulu Sierra. I am lost over the Kalahari somewhere between Gaborone and the northern sector of the Central Reserve. My fuel is critical . . . Repeat, fuel critical. Forced landing imminent. Does anyone read me?" No one answered. I suddenly felt as though I were the last survivor of some apocalypse on earth, calling into outer space with a one-in-a-billion chance of being heard and rescued by some intelligence.

My ETA for Deception Valley had come and gone. Still there was only an anonymous void below me. The left fuel gauge was faltering in the red; the right one was completely empty. I flew on, scanning ahead for a place where I could crash land with the least amount of damage to the plane and to myself.

I spotted a hint of white off to the right of my track about thirty degrees. Lake Xau! But as I flew closer, the depression taking shape in the windshield became too round, too white, to be the Lake Xau I remembered. Sure, Xau had been dry for a couple of years, but this looked too small, too much like a permanent salt pan. I couldn't see the lake bed or the Botetile River that flows into it.

If it wasn't Lake Xau, it could be Quelea Pan — in which case I was fifty miles off course to the west, deeper into the desert. It had to be one or the other. If it was Xau, I needed to turn west and fly sixty miles to get to Deception Valley; if it was Quelea, I should head east for fifty-five.

I glanced at my fuel gauges. Now both were dead red. I rolled my wings up and down and the left needle wiggled, but only slightly. There was barely enough avgas in the tank to slop around. I had to find camp or a suitable place to land immediately. I couldn't afford to waste my fuel flying closer to the pan to identify it. If I couldn't make it to camp, I wanted to get as close as possible.

If I turned west and it wasn't Xau, I would fly away from camp into a more remote part of the desert, where my chances of ever being found were nil.

There was no time left for agonizing. I turned the plane.

2

Home to the Dunes

DELIA

What aimless dreaming! The drone of the plane, the
steady sun, the long horizon, had all combined to make
me forget for a while that time moved swifter than I.

— BERYL MARKHAM

o o o

SCANNING THE HORIZON, I wondered again why Mark hadn't
flown out to look for me. I was two days late; if he'd made it safely
to camp, he would have buzzed me by now. I searched once more
for the white plane moving against the blue; but the Kalahari
Desert sky, the largest sky on earth, was empty.

Endless, barren plains — the wasted remains of Lake Xau on
the edge of the Kalahari — surrounded me. For six days I had
been driving the old Toyota Land Cruiser, burdened with supplies, from Johannesburg, across the Kalahari toward our old camp
in Deception Valley.

Mark and I had arranged to meet at camp on April 4, my
birthday. If he wasn't there when I arrived, I was to radio to the
village of Maun for an aerial search. On the other hand, if I wasn't
in camp, he was to fly along the track looking for me. My trip
across the tired scrubland had taken much longer than expected.
That Mark had not flown to look for me meant only one thing: he
had not made it to camp.

The track I was following twisted and turned across the southern tip of the dry lake bed. Driving back and forth, leaning over
the steering wheel, I looked for signs of the old track that had for
many years led us into the reserve. The plains looked so different
now after years of drought; faint tracks wandered off in odd directions and then faded altogether in the dust and drifting sand.

I climbed to the roof of the truck for a better view, squinting against the glare. A hot wind blew steadily across the wasteland. Dust devils skipped and swirled. I couldn't find a trace of the old track; either it had faded from disuse or I was lost.

There was another way into the reserve: I could drive to the top of Kedia Hill and head due west along an old cutline. It was a longer but more certain route. I turned onto the track to the hill and pressed down hard on the accelerator.

As I reached the edge of the plains, I looked back. This was where, only four years ago, a quarter of a million wildebeests had trekked for water — and died. In one day we had counted fifteen thousand dead and watched hundreds of others dying. They had migrated for several hundred miles only to find that their way to water was blocked by a great fence. For days they had plodded along the barrier until they had come to the lake plain, already overgrazed by too many starving cattle. Now it lay naked, empty, and abused. Not one wildebeest, not one cow, was in sight.

The conflict between domestic stock and wild animals had not been resolved, but we had submitted some ideas to the government that we hoped would benefit both people and wildlife. I was re-minded of how much work there was yet to do to conserve the Central Kalahari. I left the plains and headed up Kedia Hill.

Ivory-colored sand, deeper than I had ever seen, was piled high along the track and in places had drifted across the path like powdered snow. The truck's canopy and heavy load of supplies made it top-heavy; it swayed along in the spoor, leaning drunkenly from side to side. I urged it up Kedia's rocky, forested slopes and easily found the old survey track. It had been made in the early '70s by our late friend Bergie Berghoffer, who had once saved us from the desert. I felt as though he was here now, showing me the way with his cutline, which pointed like an arrow straight into the Kalahari.

Several hours later I came to the sign we'd made from wilde-beest horns to mark the boundary of the Central Kalahari Game Reserve. I stepped out of the truck for a moment to be closer to the fingers of the grass and the face of the wind. Other than the sign there was nothing but weeds and thornscrub, but we had darted the lioness Sassy just over there, under those bushes. As we put the radio collar around her neck, her three small cubs had watched from a few feet away, eyes wide with curiosity. We had known Sassy herself as a cub. If she had survived the drought, the hunters, the poachers, and the ranchers, she would be twelve now, old for a Kalahari lion. "Where are you, Sas?"

I expected Mark to zoom over the truck at any moment. He would drop down low and fly by, the belly of the airplane just above my head — one of his favorite tricks. But there was no sign of the plane.

I drove on, the truck's wheels churning steadily through the deep sand. I was glad to see that the survey ribbons left by the mining prospectors were no longer hanging in the tree. They had been shredded by the sun and blown away by the wind. The Kalahari had won that round.

Seeing fresh brown hyena tracks in the sand, I jumped out and bent down to look at them. They had been made last night by an adult moving east. I was torn between savoring every detail of my return to the Kalahari and rushing on to camp to see about Mark.

An hour later my heart began to race as I reached the crest of East Dune. I scrambled to the truck's roof and squinted under my hand, trying to see if the plane had landed at camp, nearly two miles away on the dry riverbed. The heat waves stretched and pulled the desert into distorted mirages, making it difficult to distinguish images. Even so, the broad white wings would have been visible against the sand — but the plane was not there.

Jumping to the ground, I flung open the door and drove furiously down the sand ridge. Oh God, what do I do? It had all sounded so easy to radio Maun if Mark was not here, but we had not radioed the village in four years. What if the radio didn't work? What if nobody answered?

The truck plowed on. The engine was overheating badly and complained with a deep rumbling noise — too much noise. If something was wrong with the truck, I was in bad trouble. The sound grew louder.

VAARRROOOOOM! A rush of air and thunder roared in from behind me and passed over my head. Instinctively I ducked, looking up. The belly of the plane filled the windshield as Mark skimmed ten feet above the truck. He zoomed down the dune slope and soared south toward camp. Stopping the truck, I leaned my head against the steering wheel with a rush of relief. Then I pounded it with my fists. "Damn! Where has he been? He always roars in at the last second." But I smiled. He was safe, and we were back in the Kalahari. Now I could enjoy my homecoming.

I climbed onto the roof again. I was standing in exactly the same spot from which we had first looked down on Deception Valley eleven years ago. At that time the ancient riverbed had been covered with thick, green grass and majestic herds of gemsbok and springbok. Now, stretching north and south between the dunes, the valley floor looked naked and gray, with only an occasional antelope standing in the heat. Then I noticed the faintest hint of green; only someone who had lived for years in the desert would call it green, but it was there. It had rained a few inches very recently, and the grass was struggling up through the sand. The Kalahari was neither dead nor tired, she was merely waiting for her moment to flower again.

Other people have neighborhoods that they come home to, streets with houses, familiar faces, jobs, and buildings. As I gazed down on Deception Valley, I saw my neighborhood, my home, my job, my identity, my purpose for living. Standing atop East Dune, I was looking down on my life.

Quickly I drove over the dune and across the riverbed. Mark had landed on our old strip and was rushing to greet me as I rounded Acacia Point several hundred yards from camp. I jumped out of the truck and hugged him.

"What happened? Why didn't you buzz me?" I asked.

"I almost didn't make it." Mark looked a little dazed as he recapped his flight. ". . . so I reached a point when I had to decide to go east or west. I turned west and after a few minutes recognized Hartebeest Pans. At least I knew where I was, but any second the engine was going to quit and it was still ten more minutes to the valley. When I finally landed at camp, I cut the engine and just rolled out of the cockpit onto the ground. It was a few minutes before I could even move." He had drained the tanks and measured the rest of the fuel; less than ten minutes of flying time had remained. I hugged him again and we turned toward the thorny thicket that had been our home. Camp — a lifetime in seven years. We walked back into it.

o o o

When we first decided to make this tree island our home, thousands of green branches had reached for the sky in a tangle of undergrowth. Now drought had gutted its luxuriant thicket, and its trees were gray and leafless. But here was the bush that the lions Muffin and Moffet always marked, and there was the old fireplace that had warmed our lives for more than two thousand nights. The lions of the Blue Pride had ransacked camp many times, pulling bags of flour, mealie-meal, and onions out of the trees around the kitchen boma — an open enclosure of grass and poles.

During our absence another couple had used the camp while studying desert antelope, but they had departed more than six months ago. The same faded tents lay draped across their poles, their flysheets ripped and tattered by cheeky desert storms. One side of the tent that held our lab and office had collapsed, and a small pool of rainwater from the recent shower lay bellied in its canvas. Mark planted the tent back on its poles, gingerly drew back the flaps, then with a stick chased a spitting cobra from inside. In the sleeping tent, the packing-crate bed sagged under the weight of a sodden mattress, and the tent floor was caked with mud.

The kitchen boma, with its thick, shaggy thatch roof, was still standing at the other end of camp. Inside were the cutting board Dolene had given us, the fire grate Bergie had made for us, and the blackened water kettle, scarred by the teeth of hyenas who had pirated it so often.

I looked around hopefully for the yellow-billed hornbills, those charismatic, comic birds with whom we had shared the island during every dry season. But I didn't see any. The recent light rain must have lured them back to the woodlands to mate, as it did every rainy season.

"Look who's here!" Mark exclaimed. I whirled around to see a Marico flycatcher fluttering to a branch ten inches from Mark's head. It immediately began shaking its wings, begging for something to eat. I slipped away to the cool-box in the truck and returned with a piece of cheese, one of the Marico's favorite snacks. I tossed a few bits to the ground at our feet. Without hesitation the bird swooped down, stuffed its beak with cheddar, and flew to the other side of camp.

Unloading boxes and trunks of supplies from the truck and the plane, we began the enormous job of cleaning and unpacking. Mark built a fire with some scraps from the woodpile, while I washed mud, spiderwebs, and a mouse's nest from the table in the kitchen boma. We made tea and laid out a lunch of bread, cheese, and jam on the table.

"We've got to start looking for lions right away," Mark began as soon as we sat down. April was supposed to be the end of the wet season, but according to the rain gauge only two inches of precipitation had fallen instead of the usual fourteen. Although this was enough to fill the water holes, it would evaporate in a few days. Soon the lions would be following the antelope away from the valley to their dry-season areas; we had to find and radio collar them before they left, so that we could monitor their movements with the radio receivers in the airplane and truck.

Even before our departure in 1980, the lions of the Blue Pride were already roaming over more than fifteen hundred square miles, and as much as sixty miles from their wet-season territory, in search of widely scattered prey. After four more years of drought, who could say where they were or whether they were still alive. They had led us to exciting new scientific discoveries: that they could survive indefinitely without water to drink — obtaining moisture from the fluids of their prey — and that their social behavior was different from that of other lions who lived in less harsh environments. We were anxious to continue our research for many years, and to determine how the die-off of tens of thousands of antelope had affected the lions. Their radio collars would have failed long ago; finding them would be a long shot. But if we could locate even a few, we could document not only their longevity and their ability to survive drought, but also their range sizes and the changes in pride composition during such periods.

Working feverishly that afternoon, we pulled everything out of the sleeping and office-lab tents and scrubbed the mud-caked floors. An elephant shrew with two babes clinging to her backside had to be gently evicted from her nest in the bottom drawer of the filing cabinet, and we found another snake behind the bookcase. While I continued with the cleanup, Mark prepared the darting rifle and radio collars for the lions.

Late in the afternoon, Mark carefully excavated our "wine cellar," a hole dug long ago under the thick, scraggly ziziphus trees. We had buried a few bottles in 1980, to drink on our return. The spade clanked against glass, and Mark pulled up a Nederburg Cabernet Sauvignon 1978. Sitting on the dry riverbed at the edge of camp, we watched the huge sun rest its chin on the dunes as we sipped red wine by the fire. Slowly Deception Valley faded away in the darkness.

o o o

Awakened by the distant call of jackals, we had a quick breakfast around the fire. Then, pulling the old trailer, we drove to Mid Pan to collect water. We stopped at the edge of what amounted to an oversize mud puddle with antelope droppings and algae floating on the surface. We stood for a moment, silently staring at the sludge, and we seriously considered driving out of the reserve for water. But that would take too much time — lion time. As always before, we would boil the water twenty minutes before drinking it. Squatting on the slippery mud, we scooped our cooking pots full, avoiding the animal droppings as best we could, and poured it through funnels into jerry cans. A full can weighed roughly sixty pounds, and Mark lifted each onto the trailer and emptied it into one of the drums. By the time we had collected 440 gallons, our backs and legs ached.

That evening, our second in the valley, I cooked a supper of cornbread and canned chicken stew, which we ate by lantern and candlelight in the cozy thatched boma. Then, weary but warmly satisfied with the day's work, we slid into a deep sleep in our shipping-crate bed. Not many sounds would have awakened us that night; but just as a mother never sleeps through her baby's cries, the deep rolling roar that drifted over the dunes brought us both awake at the same instant.

"Lions!"

"To the south. Quick, get a bearing."

Lion roars can carry more than five miles in the desert; the fact that we could hear them didn't mean they were close. The best way to find the big cats would be from the air, so we took off at dawn. Swooping low over the treetops, we searched for them or for vultures that might lead us to their kill. Looking in all the favorite places of the Blue Pride, we saw small herds of springbok, gemsbok, hartebeest, and giraffes. But no lions.

The next morning we heard their bellows from the south again and Mark suggested, "Look, we've heard lions to the south two nights. Let's camp down there. We'll have a better chance of finding them."

There was no track in that direction, so I drove across the dunes, making a turn just before Cheetah Tree and keeping to the east of a low ridge of sand. I chose a campsite near a clay pan where Mark could land. Seconds later he appeared, seemingly from nowhere, flew by once to check for holes, then landed. We built a fire under a lone tree overlooking the gray depression; as we ate our stew, we felt as if we were camped on the edge of a moon crater. Knowing that the lions could wake us anytime during the night, we sacked out early on the ground next to our truck, the compass by our heads.

Lion roars. Three A.M. We bolted up in our bedrolls, and Mark took a bearing on the roars. Within minutes we were driving toward them. After we'd gone a mile through the bush, we stopped to listen again. Another bellow surged across the sands, breaking over us with the resonance of a wave thundering into a sea cave. We turned the truck toward the sound and drove about two hundred yards. Mark switched on the spotlight and a medium-size acacia bush jumped to life with the reflections of eleven pairs of eyes — an adult male, three adult females, and seven cubs. They were feeding on a fresh gemsbok kill.

Mark turned off the engine, lifted his binoculars, and searched the lions for ear tags or any familiar markings we knew. But we had never seen these individuals before.

Without a minute's hesitation, all seven cubs sauntered over to investigate our truck. Only three and a half to four months old, they almost certainly had never seen a vehicle before. They walked to Mark's door and peered at him, seven small faces in a row; they smelled the tires and bumpers and crawled under the truck. Their curiosity satisfied, they began tumbling and play-fighting in a small clearing nearby, their mothers watching with bland expressions.

We sat quite near — within thirty yards — habituating them to our presence so that we could dart them that evening. By the time the sun warmed the sand, they had settled into the shade of a large bush; soon all of them, including the cubs, were asleep.

Moving to a shady spot of our own, we had a lunch of peanuts and canned fruit, then checked all of the darting equipment again. Just before sunset we drove back to the lions and found the adults feeding, while the cubs climbed all over them. Perfect. Their attention would be on the carcass, not on us, and they would be unlikely to associate the pop of the gun or the sting of the dart with our presence.

We sat very still, not making a sound, waiting for one of the lionesses to stand so that we could dart her without risk of hitting a cub. The dose intended for a three-hundred-pound lion could kill a twenty-pound youngster.

Several minutes later, one of the largest females stood and turned full flank to us. Mark loaded the dart, took aim, and squeezed the trigger. Nothing happened.

"What the . . . !?" Mark pulled the gun back through the window and thumbed the safety "on." The gun fired and the dart flew out the window into the bushes. Mark cocked the unloaded gun and pulled the trigger; it didn't fire. When he clicked the safety on, it did. Whatever the problem was, we had no time to fix it. The more the lions gorged themselves, the more drug it would take to sedate them. Mark would use the safety to fire the rifle. He took aim again as one of the females stood up; but as he did, a cub crawled under her neck.

Mark waited a few seconds, his cheek against the gun stock, while the cub moved past the lioness. He thumbed the safety on and the gun fired with a muted pop. Just as the dart lobbed out of the barrel, another cub stepped from under the female's belly. We watched helplessly as the dart arched lower and lower, striking the cub in the flank. He squealed, spun around, and stumbled off into the thick bush.

"Good God!" I cried.

"Where did it come from? The cub I was watching moved off!"

We had darted lions and other carnivores more than a hundred times, and nothing like this had ever happened before.

"Should we go after it?" I asked.

"There's nothing we can do. That cub doesn't have a chance. Let's concentrate on getting collars on the adults," Mark said. "First, we have to see about this damned gun." As saddened as we were about the cub, I knew Mark was right. We drove off about four hundred yards and while I held the flashlight, Mark repaired the gun on the hood of the truck.

After driving back to the lions, Mark darted the female that he'd missed, then the male, whose golden mane tinged with black was one of the most beautiful we had ever seen. The two darted lions moved off into the bush where, seventeen minutes later, they slumped down under the influence of the drug.

I put salve in the male's eyes, while Mark injected him with antibiotic. We collared and measured both him and the female, keeping an eye out for the two undarted lionesses, who had disappeared into the bush.

After checking the breathing and pulse of the darted lions, we moved off a hundred yards and sat for an hour until they were yawning and stretching, fully recovered. Then as Mark began driving back to the plane, I saw a small lump that looked like a rag on the cold sand.

"Mark! It's the cub!" We jumped from the truck and walked to the infant.

I watched for the adults as Mark squatted beside the cub, slipping his fingers between its front leg and chest and feeling for a pulse. The cub's body was already cool. Several seconds went by, then Mark felt a little blurb of pressure beneath his fingertips. Pushing his fingers deeper into the fur, he detected a single subtle pulse.

While I rushed to the truck to get the drug boxes, Mark massaged the cub, trying to stimulate his heart. He gave the little lion an intravenous injection of Doprim, a respiratory stimulant, and a massive intramuscular injection of antibiotic. Within minutes the cub's pulse was stronger but he was still hypothermic. We gently laid a canvas tarp over him, covering everything but his face, and heaped a pyramid of sand over his body for extra insulation. I stroked his muzzle once more, then carried the drug boxes back to the Cruiser.

Mark was making a fifteen-second count of the cub's pulse when we heard a loud splintering crash and a growl. Mark whirled and saw a lioness crashing through an acacia bush forty yards away. As he sprinted for the truck, I grabbed the spotlight and flicked it on, trying to dazzle the lioness. But the light shone directly into Mark's eyes, blinding him instead. Holding his arm over his face, he staggered forward. I dipped the light so he could see. The lioness stopped at the cub, sniffing it briefly. Then she bounded over it and ran toward Mark, her big feet drumming on the sand.

With one hand I swept the darting equipment from the front seat onto the floor and slid into the driver's seat, ready to start the engine. Once more I held the light on the charging lioness.

Blinded again, Mark slammed into the side of the front fender of the truck and stumbled back. He reached up for a handhold, tried vaulting onto the Toyota's hood, but missed and fell to the ground.

Swinging the light back and forth over the lioness' eyes, I opened the door, screaming "Get in! Get in!" Jumping to his feet, Mark fumbled frantically for the door. Finally he found it and dived in, crawling over me to the other side of the front seat.

The lioness broke off her charge only eight yards away. Her tail flicking, she walked back to her cub, sniffed its head and the sand over its body, then strolled back to the gemsbok carcass. The other lions were feeding again and had ignored the commotion. Both of us slumped against the back of the seat and breathed deeply. We would name that lioness Stormy.

We drove the truck to a lonchicarpus tree several hundred yards from the newly collared lions. I dug out some baked beans and we ate them cold out of the can. It was after midnight; we had been working almost continuously for twenty-two hours. My eyes felt as though they had sand in them, and Mark's shins and knees throbbed from hitting the truck.

We reeked of lions, the way a cowboy smells of his horse — a dank, earthy, not altogether unpleasant odor. But because carnivores sometimes carry echinococci, parasites that can infect the human brain, we splashed some cold water and disinfectant into a basin and washed thoroughly. Too tired to drive to the plane, we laid our foam mats and sleeping bags in the back of the Land Cruiser and crawled inside. Toolboxes at our feet, jerry cans at our heads, and the back door standing ajar, we slept.

We opened our eyes to a crisp Kalahari dawn, sunlight streaming over the dunes. After a quick breakfast we drove back to the dune crest to check on the lions. As soon as Mark switched off the truck, the collared female yawned deeply and began licking her front paw. We named her Sage and the male Sunrise. Stormy watched us carefully for a few minutes. Then even she began to nod and snooze in the morning sun, apparently at ease with us. Saucy, who along with Stormy had avoided being collared last night, slept with her head on Stormy's flank. Sitting among the heap of lions, we were pleased to have a new pride, though disappointed that we had not found the old one.

We drove to the spot where we had left the darted infant. The small pyramid of sand was flat; the cub and tarp were gone. I was hopeful that he had recovered, but Mark pointed out that he wouldn't have taken the tarp. "A hyena or a jackal may have got him," he said.

I drove Mark back to the plane and he took off from the clay depression, the plane bouncing over the rough ground. He flew over, checking the radio collars from the air, then headed on to camp.

Returning to the lions, I circled them in the truck, searching for the darted cub but finding no sign of him. I parked under a shade tree and began copying the notes I'd made the previous night. Dozing in the hot truck, I lifted my head from time to time to check on the lions. Now and then they shifted their position for better shade, and so did I.

Just before sunset one of the cubs bounded from the thicket, chased by another. Then three ran full speed across the clearing and behind a bush. Two dashed into the open, tearing and pulling on the dead gemsbok's tail. Were they the first pair, or two new cubs? Now four tumbled through the grass in a mock hunt, and one of them dashed out of view as two others pulled a piece of canvas. They were playing with the tarp! Two more cubs bounded into the scene. Seven! There were seven cubs! The one we darted was okay; in fact, I could see no difference between his behavior and that of the others. I counted again, just to be sure. Seven. I smiled.

The adults also moved into the clearing and lay in the last rays of sunshine, while the cubs attacked their ears, muzzles, and tails. As Stormy began walking south, Sage stood, stretched, and followed. The cubs trotted to catch up, and finally Saucy and Sunrise trundled after the others in a long, rambling line. As the sun set, I watched their golden bodies glide through the blond grass until they disappeared. Then I headed for camp to tell Mark the good news.

o o o

CRASH! A tin trunk full of canned food hit the ground in the kitchen boma. I looked at my watch; it was 5:30 A.M. We jumped from bed and pulled on our jeans. We had been in the Kalahari for almost six weeks and had darted eight lions in three prides. But still we had not found the Blue Pride.

Pushing back the flap of our tent, we peered out and saw the female lions romping around the campfire. Sage was dragging the ax handle in her mouth, while Stormy pawed at its head. Saucy was standing inside the grass boma, sniffing the pots on top of the table. We tiptoed down the path through camp to get a better look.

Two of the still unpacked boxes of food supplies lay on their sides with tins of oatmeal and powdered milk scattered around the campfire. Saucy chomped into a pot with her teeth and, holding it over her nose, pranced from the boma. The others chased her.

Their bellies were high and tight to their spines, a sign that they had not fed for several days. We had been following them for the previous five nights and they had not made a kill. Their cubs were nowhere to be seen. Seven cubs were too many for inexperienced mothers in these dry times; under such conditions the young are often abandoned.

Then my eyes met those of a fourth lioness, standing just beyond the trees of camp. We stared at each other for long seconds. She was old; her back sagged, and her belly hung low. For some reason she did not join in the play. Was she too old for it? Or had she played this game too many times before? We looked for ear tags or scars; there were none.

"Hey! Come on, that's enough," Mark called as Stormy stuck her head into the supply tent. He clapped his hands loudly and the lioness backed up, looked us over, then ambled back to the kitchen, where she grabbed a dish towel and ran out of camp. The other two followed and they chased one another around the plane. After a while the three young lionesses calmed down and, with the old one, walked north along the track. We grabbed some peanut butter and crackers for breakfast, got into the truck, and followed. They paused on the other side of Acacia Point, then broke into a trot to greet Sunrise, the newly collared male, who was swaggering from the bushes of East Dune onto the dry riverbed. After rubbing sinuously along his mane and body, the pride continued north toward the water hole on Mid Pan.

At the water's edge they lay flank to flank and drank for several minutes, their lapping tongues reflected in the water. Sunrise lifted his tail and scent-marked a thicket — the same thicket the Blue Pride lions had always sprayed when they passed the water hole. As the pride settled down under a shade tree at the base of East Dune, we returned to camp to prepare the darting equipment. Tonight we must collar Stormy, Saucy, and the old lioness.

When we returned in the late afternoon, Sunrise was feeding on a twenty-five-pound steenbok in the tall grass of East Dune. He had probably taken it from the lionesses moments earlier. Fifty yards away Saucy and the old lioness were feeding on a freshly killed gemsbok. But Stormy, Sage, and their cubs — the ones who most needed the meat — were nowhere in sight.

Mark darted Saucy and the old lioness with the sagging back, and they wandered away from the carcass into the bush, where we could treat them without disturbing Sunrise. Working quickly, we collared Saucy first. Then Mark nudged the old lioness gently with his foot to be sure she was properly sedated. Crouching beside her, he pushed back the hair on her left ear, uncovering a black plastic pin and a tiny piece of yellow plastic — the remains of an old ear tag.

I thumbed quickly through the identification cards of all the lions we had known. Blue — blue tag in right ear; Sassy — red tag in right ear; Happy — yellow tag in left ear . . .

"Mark, it's Happy!" We sat down next to the old lioness and stroked her. As a young female of the Springbok Pan Pride, she had invaded the Blue Pride's territory, won acceptance from its resident females, raided our camp with them, sat with us in the moonlight, slept near us, and — finally — had beguiled us, as she had the males Muffin and Moffet. We had spent hundreds of hours with her as we tried to understand the ways of desert lions. She had often swapped prides and males and had wandered away from Deception Valley many times, but she had always come back. Now she was a matron who had made it through one of the worst droughts the area had known.

We gave Happy a new yellow tag and radio collar, measured her body, and took pictures of her worn teeth. During all of this we touched her more than was necessary. By the time we finished, I had memorized her face.

Happy lifted her head slightly and looked around. Reluctantly, we backed away to the safety of the truck to watch her recover. When we were satisfied that both darted lionesses were recovering well, we drove the truck around some bushes to the gemsbok carcass. Sunrise was sleeping a short distance away, his belly round and heavy with meat. Feeding on the carcass were Sage, Stormy, and the seven cubs.

The cubs were already very full, their small bellies bulging like melons. They pulled at the fresh meat for a few more minutes, and then three of them plopped down and fell asleep. The other four tumbled around the grassy clearing, spending most of their time bouncing on Sunrise's expansive stomach. We were almost too elated to leave, but the deep yawns of the lions were contagious. We headed toward camp.

The drying grasses of the dunes glowed in the light of the full moon, which was so bright in the cloudless sky that we drove without headlights across the valley floor. But as we stepped from the truck at camp, the light began to dim, giving the desert a shimmering blue-gray cast. We looked up to see that the earth's shadow was stalking the moon; a full lunar eclipse was under way.

Pulling the foam mattress and sleeping bags out of our tent, we laid them on the ancient riverbed, next to several *Acacia tortillas* trees at the edge of camp. Their twisted, thorny branches had somehow spited the drought by producing flowers, then corkscrew pods full of seeds. From our spot we could see five miles along the valley and watch this secret desert drama before drifting off to sleep. Slowly the earth drew its shadow across the face of the moon, and the Kalahari grew dark and silent.

Moments later the stillness was broken by the clopping of heavy hooves. Three stately giraffes glided into view above us, silhouettes against the darkening sky. Apparently they had not noticed the two lumps on the ground, and it was too late for us to move without frightening them. We lay still, swaddled in our sleeping bags, literally at their feet. They spread out a few yards away, browsing the pods from the acacias. Lying almost under the giraffes' bellies, in this silky light we felt as if we were being absorbed into the desert.

The next morning, May 13, the entire eastern horizon was lined with patchwork clouds, blushing deep pink at first, then transforming into a quilt of gold when the sun found the dunes east of Deception Valley. As we ate pancakes around the campfire, I looked out over the valley; all seemed well with the Kalahari. We had heard reports — perhaps only rumors — that the government was planning to turn the lower two-thirds of the reserve into cattle ranches. Even though thousands of wildebeest had already died along the fences, there might still be time to resolve the conflict between cattle and wild animals. Knowing that Happy had lived through the drought gave us renewed hope that the reserve itself would survive.

Mark walked to the office tent for the radio schedule with Sue Carver, our contact in Maun, more than a hundred miles to the north. Meanwhile, I cleared away our dishes and fed the Marico flycatchers.

"Hello, Mark. I have an important message for you," I heard Sue say, her voice crackling.

"Hi, Sue. I'm ready to copy. Go ahead."

"It's from the Immigration Department. They say that your research permit has been denied and that you are to report to the Immigration Department in Gaborone immediately. Repeat, you must report to the Immigration Department immediately."

3
Against the Wind

MARK

> Some things just
> don't go on. some circles
> come undone. some sparrows
>
> fall. sometimes sorrow,
> in spite of resolution,
>
> enters in.
>
> — PAULA GUNN ALLEN

o o o

"ARE WE BEING JAILED?" I asked. His face set in stone, the immigration officer said nothing and continued rolling my fingers over an ink pad, then pressing them on white cards, one each for the military, police, and immigration authorities. The day before, we had flown to Gaborone, the capital of Botswana, and this morning had been detained at immigration headquarters.

"Please, I would like to telephone the U.S. embassy," I said.

"No."

He finished taking my fingerprints, and I watched as he began with Delia. Another man took my elbow firmly and started to lead me away.

"Wait, please, don't separate us . . . ," Delia pleaded in a small voice. Pulling my arm free, I walked back to her side and said to the second man, "Just wait till he's through, okay?" He released my arm. When they had finished taking Delia's prints, another officer pushed two forms across the desk toward us. "You must sign these."

"*Declaration of Status as Prohibited Immigrants (PI)*" leaped off the page. They were throwing us out of the country! Before I could read any further, he snatched the forms away.

I swallowed hard and asked politely, "What does it mean if we sign these forms? I would like to see an attorney first."

The officer stood up abruptly and strode across the room. When he returned, he was followed by a giant of a man six and a half feet tall, weighing about two hundred fifty pounds. The big man glared down at me. "What is your problem?" he rumbled. "Sign, and then you can go about your business."

"I'm sorry, but we can't sign this without reading it first." I tried to explain. "When we were intercepted and brought here, we were on our way to the permanent secretary to the president with this letter appealing the denial of our research permits." I held up the envelope.

"After you sign, you can go to the president's office or wherever you want. And you can appeal the PI ruling. But you must sign these forms now!" He slapped the sheets onto the desk in front of us. I started to protest again, but he leaned over until his face was inches from mine.

"Sign, or the law will take its course. Do you understand what that means?"

I looked at Delia and put my signature on the PI notice. She did the same. The instant I lifted my pen from the form, he jabbed his finger at a paragraph near the end of the page.

"You will note in this subsection," he said, "that when the declaration is by presidential decree, as in this case, there is no right of appeal."

"But you just told us . . ."

"If you read the form, you will see that what I say is true and that I have no choice in the matter," he cut me off.

For the first time we were allowed to read the document we had signed. It stated that the president himself had ordered our deportation, that we could not appeal his decision, and that no reason need be given for it.

The big man led us to a small room where he stood facing us, his back to the wall, arms folded across his barrel chest. A uniformed policeman was seated behind a desk.

"As of this moment you are in Botswana illegally," the policeman said.

"But why?" Delia spoke up. "We've done nothing wrong. What are we charged with?"

"I'm just a cog in the machine. And even if I knew, I couldn't tell you. You must be out of the country by five o'clock. Do you know what the law expects of you now?"

It was already two-thirty. They were giving us only two and a half hours to get to the hotel, pack, go to the airport, plan our flight, check through customs and immigration, preflight the airplane, and take off.

"Look, please, we have thousands of dollars worth of equipment at our camp," I pointed out. "We need time to go back there and dispose of it. And what about the weather? We're in a small plane. Clouds are building up. It may not be safe for us to take off."

He leaned toward us, scowling. "I say again, if you are not out of this country by 5:00 P.M. the law will take its course! Do you understand?"

We rushed back to the hotel, threw our clothes into our suitcase, and stopped briefly at the American embassy to report what had happened. At four-thirty-five we hailed a cab to the airport. Fifteen minutes after takeoff, the Limpopo River slipped by below us. On May 15, 1985, as we left Botswana's airspace, we passed from wild, innocent Africa with its sweeping savannas of plains game and wide rivers of sand into a new era of confusion, turbulence, uncertainty, and danger. At the Limpopo we flew into a strong head wind.

4

Beyond Deception

DELIA

The woods where the weird shadows slant,
The stillness, the moonlight, the mystery,
I've bade 'em good-by — but I can't.
    ....

There are valleys unpeopled and still;
There's a land — oh, it beckons and beckons,
And I want to go back — and I will.

— ROBERT SERVICE

o o o

STANDING IN THE MIDDLE of a field, five hundred miles south
of Deception Valley, I looked up at the moon. At this moment the
same full moon was hanging over the desert, and I wished that I
could somehow see the reflection of the dunes and the old riverbed
in its face. Was Happy still with Stormy and Sunrise? How were
Saucy's cubs, and Sage's? Like the wildebeest, we could no longer
move freely into the desert; we were another casualty of the fences.

Botswana gave no official reason for expelling us. Informally,
the ambassador to Washington told us that his president, Quett
Masire, had been angered by our reports on the fences that
blocked Kalahari migrations and resulted in the deaths of
hundreds of thousands of desert antelope. But our accounts were
accurate, and we believed it our responsibility to report the disastrous effects of these fences on wildlife (see Appendix A). Later
another Botswana official confided that we were really deported
because powerful ranchers-cum-politicians had wanted to establish their private cattle ranches in the Central Kalahari Game
Reserve, and they knew we would speak out against their scheme.
One of the longest-running scientific studies of lions and other
carnivores in the wild and one of the largest wildlife protectorates

in the world were both being tossed aside for the financial benefit of a few people.

Writing appeals, reports, letters, we tried frantically to get back to the desert. Many people, including U.S. congressmen from both parties, Atlanta's Mayor Andrew Young, and then Vice President George Bush, requested that Botswana's officials allow us to return. But they refused to discuss the issue — even to respond to the vice president of the United States. Months passed.

Time was being wasted; lion time, hyena time, conservation time, a lifetime, it seemed. We wrote more letters, made more phone calls. But there was no answer from Botswana.

After eight months Mark accepted the fact that we were banished from the Kalahari and wisely decided that we should search for a new wilderness to study. But hope still stalked me. Even after all this time, I believed a letter would arrive or a telephone would ring with the message that we had been misunderstood, that Botswana had relented and would allow us to go back to Deception Valley.

I gazed at the farm cottage where we were staying, outside Johannesburg. All but smothered in flowering vines, it was another home that had opened its arms to us, another wonderful family, another friendly dog. We had been living out of suitcases for months, always in someone else's back room or guest cottage — from California to Johannesburg — and had had so many different addresses that our mail rarely caught up with us. A trail of unanswered letters and spoiled dogs lay behind us.

One day I noticed a tab of paper pinned on a friend's bulletin board. Amid Gary Larson cartoons and holiday photos was a quote by Alexander Graham Bell, "Sometimes we stare so long at the door that has been closed to us, we do not see the many doors that are open." I read it twice, a third time, then walked to where Mark was writing and said, "It's time to find another Deception Valley." We would go in search of a new wilderness, and with a new idea.

For years we had believed that, at least in some places, wildlife can be more beneficial to a country and its people than exotic agricultural schemes. Too often aid and development agencies sweep aside the valuable natural resources in an area so they can get on with "real" development. They chop down lush forests and kill off wildlife, only to plant crops that deplete the soil of nutrients and yield poorly; they irrigate arid lands until they are sterilized by mineral salts; they overgraze grasslands, turning them to deserts.

This is what had gone wrong in Botswana. The Kalahari was teeming with wildlife whose migrations had adapted them to long droughts and sparse grasslands. These animals could be used for tourism, game ranching, safari hunting, and other schemes that would bring revenue to a large number of local residents, including Bushmen. Instead, the World Bank, the European common market countries, and the Botswana Development Corporation wanted to replace wildlife with cattle. Large-scale commercial ranchers in the Kalahari had already killed off hundreds of thousands of wild animals, overgrazed the desert, and depleted the water from fossilized aquifers. They had left a wasteland that was good for neither wild nor domestic stock.

In most places on earth, Nature long ago figured out what works best, and where. Often the best improvement humans can make is to leave everything alone. Nowhere is this more true than in marginal lands. The least we can do — before we chop down trees or build long fences — is watch for a while, to see if we can make the natural resources work for us in a sustainable way. Perhaps if local people who live near national parks could benefit directly from them, for example through tourism, they would recognize the economic value of wild animals and work to conserve them.

It was an idea worth exploring. But first we had to find a place.

o o o

Standing over a map of Africa, we eliminated one country after another. The continent seemed to come apart in pieces: Angola and Mozambique were torn with civil wars; Namibia was under attack from SWAPO (the South West Africa People's Organization), and human overpopulation had just about finished off the wildlife in western Africa. Sudan was out: the Frankfurt Zoological Society, our sponsors, had recently lost a camp to the Sudanese Liberation Army, which had kidnapped the staff members and held them for ransom. As Mark's hand swept across the map, wild Africa seemed to shrink before our eyes.

The region most likely to have large wilderness areas was tucked under the shoulder of the continent in Zimbabwe, Zambia, Zaire, and Tanzania. We began to outline the hundreds of details for a three-thousand-mile expedition north from South Africa through these countries. Mark would fly the Frankfurt Zoological Society airplane and I would drive the truck with its trailer; we would meet along the way at potential sites. We contacted the American embassies to find out where we could get aviation and diesel fuel, and where it would be safe to land without fear of partisans or bandits. We bought, labeled, and packed camping gear, foodstuffs, and scores of spare airplane and truck parts that would not be available on our route.

Finally, a year after being expelled from Botswana, we were almost ready to depart. Several travelers had recently been murdered along the main roads through Zimbabwe and Zambia, however, and the American embassies in those countries had issued travel warnings to U.S. citizens. So that I would not have to drive alone, Mark prepared to fly our plane to Lusaka, Zambia. He would leave it at the airport, then return to Johannesburg on a commercial flight so that we could ride together to Zambia.

On the morning of May 19, 1986, Mark drove to Lanseria Airport just north of Johannesburg, where our Cessna was hangared, to make his flight to Lusaka. Standing in the open door of the plane, he was loading his duffel bag and flight case when a man rushed up behind him and panted, "Excuse me, I believe you're flying to Lusaka?"

"That's right," Mark answered.

"Any chance of a lift?" the man asked hopefully.

"No problem," Mark assured him. "What's the hurry?"

"Haven't you heard? The South Africans bombed Lusaka this morning. And they hit ANC [African National Congress] hideouts in Botswana and Zimbabwe!"

"Lusaka! Are you sure?"

"Yeah. I'm a UPI reporter; I've got to get there quick." Mark stared at him for a moment, then said, "I don't know about you, pal, but I'm not flying to Lusaka today."

Our Cessna 180K was the same model that South African defense forces used for reconnaissance flights, and it still bore its South African registry. After canceling his flight, Mark returned to the cottage where we were staying. We sat at the table reading the latest news releases: "South African Defense Force hits three capitals in the biggest operation so far launched against ANC targets."

Our plan not only called for Mark to fly into Lusaka, which had just been bombed, but to fly the length of Zimbabwe, which had also been attacked and was known to have antiaircraft guns and surface-to-air (SAM) missiles. It would be foolhardy to fly over these territories with a South African–registered plane. But since we had purchased the plane in South Africa, by international law it had to retain that registry until we officially imported it into another nation and reregistered it. And that we could not do until we had settled in a new country.

"I'll give it a week; maybe things will cool down," Mark said, looking over his paper at me.

"Why not fly over Botswana instead of Zimbabwe? There are fewer antiaircraft batteries and missile installations in Botswana," I suggested.

"Much longer flight, and I might not have enough fuel. Don't worry, Roy told me how to avoid missiles."

Roy Liebenberg, a former military pilot, had taught Mark how to fly. They still kept in touch. His most recent bit of advice: "Stay real low so they can't get a lock on you. If you see a launch, climb straight for the sun until the missile is right behind you. Then chop the power, break hard right or left, and dive for the ground." Roy had also warned Mark about flying into Lusaka International Airport. Understandably, the Zambians were somewhat trigger-happy since the South African raid, and apparently they had accidently shot down two of their own military planes.

"I can't believe we're having this conversation," I said. Mark shook his paper and went back to his reading.

After several days, news of the attacks died away, and it became apparent that South African forces had made surgical strikes against ANC headquarters rather than more general attacks against the Botswana, Zimbabwe, and Zambian governments. Mark phoned the American embassy in Lusaka, and although the official with whom he spoke understandably did not want to give any guarantees, he said that life in Lusaka was going on "pretty much as normal and a flight should be fine."

On May 26, a week after the raids, Mark filed an official flight plan informing the Lusaka control tower that he would arrive at their field that evening. The plan went through by telex and no specific instructions or warnings were issued, so Mark took off. He flew checkpoint to checkpoint over Botswana and Zimbabwe for five hours. Darkness had rolled in beneath him by the time he was over the eastern end of Lake Kariba, the outline of its shore only faintly evident from the cooking fires of remote villages. Lusaka was still fifty minutes away.

When he approached Lusaka airspace, the controller did not answer his radio calls. Mark tried again and again. No response. He could not know it, but soldiers were manning antiaircraft guns at the end of the runway. The tower controller had alerted them to the approach of an unauthorized South African–registered plane. Cranking the guns around, they fixed their sights on the Cessna.

Even though Mark had not heard from the controller, he had no choice but to land; his fuel was almost gone. Approaching from the east, he lined up on the main runway and flew directly toward the antiaircraft battery.

When the plane was off the end of the runway, the gunners began to finger their triggers. Suddenly a Land Rover roared to a stop, and a colonel in the Zambian air force jumped out, yelling and waving his arms as he ran toward the gun battery. Seconds later Mark glided over the end of the runway and touched down.

Standing about a hundred fifty yards away in its own pool of harsh light, the terminal building looked deserted. Mark climbed out, stretched, unloaded his luggage, and began tying down the plane for the night. All at once, six soldiers stormed toward him from the building, their Kalashnikov (AK-47) rifles leveled at his stomach. "Halt! Do not move!"

Two of the soldiers grabbed Mark by the arms and steered him into the building and to a room with a faded blue "Police" sign over the door. The others followed with their AKs still leveled at the prisoner. They sat Mark on the bench and stood back, waiting.

Soon the colonel strode into the room, pulled a chair in front of Mark, and sat down facing him. For seven hours he grilled Mark on who he was and what he was doing in Lusaka. Fortunately, Mark had a briefcase full of introductory letters from the U.S. embassy, research permits, customs clearances, and a copy of his official flight plan. Finally, at 3:30 A.M. the colonel shook his index finger in Mark's face. "I was called to the antiaircraft battery as you were approaching the field. My men wanted to open fire and shoot you down. If I hadn't been there, you would be dead right now."

The next day Mark returned to Johannesburg on a commercial flight, wondering where a biologist fits on this tormented continent.

o o o

Our trunks were packed, and preparations for the journey north were complete. We were having supper in the A-frame cottage in Johannesburg, on our last night before departure, when the phone rang. Kevin Gill, our longtime friend, confidant, and legal counsel, was on the line. Some of our mail was still being delivered to his home, where we often stayed, and he told me that we had received an official letter from the government of Botswana. This was their first communication since they had deported us a year ago.

"Would you like me to read it to you, Delia?"

"I guess so, Kevin," I said.

There was a brief silence and the sound of shuffling papers. "Yes, it's what I thought." The letter from Mr. Festes Mochae, personal secretary to the president, was short and to the point: "The president has carefully considered all these appeals and has decided to lift your status as Prohibited Immigrants."

I muttered a word of thanks to Kevin, hung up, and ran to Mark. For months we had tried to get a reversal of the deportation. We had finally given up and set our sights on a new goal. And now we could go back to the Kalahari. We stared at each other in a confusion of emotions.

We had been cleared of any wrongdoing, but a lot of international pressure — from the United States and Europe — had been brought to bear on Botswana for deporting us simply because we reported an environmental problem. Other scientists had visited the desert and confirmed that our reports of the dying wildebeest were accurate. People were outraged about the fences; we were no longer the issue. Still, because of all the controversy we knew that we would not be welcome in Botswana at this time. We had no choice but to carry on with our plan to search for a new location. One day we would go back to Deception Valley to look again for a lioness named Happy. But that would be much later.

o o o

"Toyota Spares." "Airplane Spares." "Everyday Tools." "Everyday Food." "Food Stores." "Cooking Kit." "Bedding and Mosquito Nets." "Lanterns and Accessories." "Reference Books and Maps." "Cameras." "First Aid." "Mark's Clothes." "Delia's Clothes." Carefully labeled heavy trunks filled the truck, along with a mattress, folding chairs and tables, a chuck box, and two jerry cans of water. In the trailer were five drums of aviation gas, a drum of diesel fuel, three spare tires, a pump, a tent, shovels, axes, two high-lift jacks, ropes, and tarps. Driving our tired old Land Cruiser and worn-out trailer, their homemade bodies patched and repatched with scrap steel, we inched our way up Africa. None of the rusty blue trunks of supplies gave a clue to the dreams and the hopes that were packed inside.

During one portion of our journey through Zimbabwe, we were a hundred miles directly east of the Kalahari. Low, dark clouds stretched endlessly across the sky to the west, and we thought that perhaps rain was falling on the desert. Maybe the long drought had ended; maybe Happy, Sage, and Stormy would at last get a taste of water. On June 2, 1986, we crossed the Zambezi River, and headed north toward another season.

o o o

"We thought we would try Liuwa Plain National Park next," Mark said to Gilson Kaweche, chief research officer for Zambia's national parks. We had just spent five weeks exploring Kafue National Park in east central Zambia, often camping in places that had not seen a human in more than twenty years. Kafue was big and beautiful — the size of Wales — but hordes of commercial poachers were exterminating all the wildlife there. The park and its problems were too big for our resources.

Kaweche shifted uneasily in his chair at our mention of Liuwa Plain. "Ah, well, I'm sorry to say that security is a problem there, because of the UNITA [National Union for the Total Independence of Angola] rebels. Anyway, most of the animals in that park were shot long ago."

"I guess in that case we could try West Lungu Park first."

Kaweche's brow wrinkled as he concentrated on the doodle he was drawing. "Yes, but unfortunately in Lungu you will have a similar problem with security: some Zairian smugglers have been laying landmines along the roads. It would be highly risky for you to go there. I doubt my government would permit it."

"And Sioma Park, down in the southwest? What is the situation there?" I asked.

"Well, again it's the security problem. Sioma is right on the Caprivi Strip, which is South African territory. Freedom fighters from Angola cross the strip into Botswana on their way to South Africa. The South African army is trying to stop them. It would be unsafe for you to work there."

"How about Blue Lagoon, on the Kafue River . . . ?"

"I'm afraid the army has taken over that national park."

"How can the army take over a national park?"

"The military can do anything it wants." He chuckled.

One by one, we asked about the nineteen national parks shown on the maps of Zambia. Most were parks on paper only.

"We'll have to try Tanzania," Mark said to me. Our permits to look for a research site there had not yet been approved, so we would have to enter the country as tourists. If we found a suitable place for our research, we would request permission to stay.

I glanced up again at the map on the wall, my eyes traveling along the route that would take us through Zambia to Tanzania. More than four hundred miles up the road from Lusaka was another national park. "What about North Luangwa?" I asked.

"I'm sad to say that we have about written off the North Park," he replied. "It is just too remote and inaccessible to protect. No one goes to North Luangwa, so we have no idea what's happening there. I've never seen it myself, but I've heard it is a beautiful place."

"Anything wrong with our stopping there to take a look, on our way to Tanzania?"

"No," he said, "just give us a report on what you find." Gilson went on to warn us that this was not a "national park" in the American sense. There were no tourist facilities, no roads, and no one living in the park — not even game scouts. It was a 2400-square-mile tract of raw wilderness. Seasonal flooding of its many rivers made it impassible in the rainy season. The sectional map that Gilson spread over his desk gave no hint of even a track leading into the valley. Remote, rugged, and inaccessible — North Luangwa sounded like our kind of place.

After thanking Gilson, we visited Norman Carr, an old poacher-cum-game-ranger-cum-tour-operator, who in his eighty-odd years has come to know the valley better than any other African. Carr leads walking safaris in South Luangwa National Park, and his tough hide and infinite knowledge of trees, birds, and mammals are testimony to his expertise.

"Forget it. North Luangwa is impossible. You'll have a bloody time getting around in the dry season because of all the deep ravines and sand," he said. "And you can't drive around in the wet season because of all the mud. Those flash floods — they'll wash your truck away, even your camp."

Maybe. But we were determined to see for ourselves. Besides, if North Luangwa was not the wilderness we longed for, where else could we go?

PART TWO °°°

A Season for Change

Prologue

MARK

THE SUN SINKS SLOWLY behind the mountains of the scarp as One Tusk, the elephant matriarch, steps cautiously from the forest along the Mwaleshi River in Zambia. Holding her trunk aloft, she searches the wind for danger. She is thirsty, as are the four young females in her family, one with an infant that gently presses his head into his mother's flank. Weeks earlier the rains tapered off, and by now most of the water holes away from the rivers are liquid mud. The elephants have come a long way since yesterday without drinking. They hurry forward, eager to cool themselves in the river after the heat of the day. But the matriarch holds them back, perhaps remembering an earlier time when poachers had chosen such a place for their ambush. She waits, her mouth dry with fear and drought, as the little calf nuzzles her mother's withered breast.

At that same moment, in Mwamfushi Village, far upstream of the elephants, another mother holds a crying infant to her flaccid breast. The stingy rains have turned the millet and maize to yellow, shriveled weeds. There will be starvation in the village this year unless the men go hunting in the park — unless Musakanya, her young husband, goes poaching.

For the past two weeks the family has lived on little more than *n'shima*, a paste of boiled maize meal dipped in a gravy made with beans. They crave meat, and Musakanya knows where to get it. He shoulders his rifle and walks down a dusty footpath that sixty miles later will end in the North Luangwa National Park. At the edge of his village, under the tree where they always meet on these expeditions, he joins Bwalya Muchisa and Chanda Seven, two friends who will poach for more than meat; they are going for ivory.

5

Into the Rift

MARK

Wilderness is not dependent upon a vast, unsettled tract
of land. Rather, it is a quality of awareness, an openness
to the light, to the seasons, and to nature's perpetual re-
newal.

— JOHN ELDER

o o o

SEVERAL DAYS AFTER our meeting with Gilson Kaweche at the
National Parks headquarters, Delia drives and I fly the four
hundred miles from Lusaka to Mpika. We sleep a cold, windy July
night on the airstrip. At sunrise the next morning we take off down
the runway into a strong wind. Zulu Sierra rises like a kite over a
forested hill and within five minutes the last thatched hut has
slipped from view below us. Soon after, the forest floor begins to
show its first ripples and rills — the effects of titanic stresses along
the Rift Valley. Two massive tectonic plates, one on each side of
this gigantic trench, are drifting apart, tearing Africa in two. Taller
mountains loom ahead, like sentinels guarding the valley. We
climb over them, and fly along great ridges of rock, then over deep
canyons, partly hidden by tropical trees and luxuriant sprays of
bamboo, that seem to plunge away to the very center of the earth.
Rushing rivers and waterfalls cascade over walls of sheer granite.

Suddenly a huge jawbone of rock runs northeast-southwest
across our track as far as we can see — the Muchinga Escarpment,
the western wall of the great Rift. Massive blunted mountains are
rooted in this jaw like mammoth crooked molars, and whitewater
streams burst between them, coursing untamed down the apron of
the scarp and into the valley. According to our charts, these rivers
— the Lufishi, Mwaleshi, Lufwashi, Mulandashi, and Munyamadzi — stream off the scarp to join the larger, wilder Luangwa

River. Flowing along the eastern border of the park, the Luangwa wanders to and fro over the valley floor, spreading the rich alluvium its tributaries have eroded from the plateau to the west beyond the Muchinga. It flows from Tanzania into Zambia, on to the Zambezi, and thence to the Indian Ocean between Mozambique and Malawi.

As we fly between two rounded cusps of the escarpment, the earth below us disappears, just drops away, leaving nothing but white haze under the plane. I pull back the throttle and we descend more than three thousand feet through the murk to the valley floor. Leaning forward, I watch for any peaks that might reach up to gut the belly of our plane, while Delia tries to spot a topographical feature that will tell us where we are.

Minutes later, the serpentine shape of a sandy river gradually emerges from the haze, as though we are regaining consciousness. Flying low, we follow the Lufwashi's tortuous route as it cuts its way out of the mountains along a ridge peppered with herds of sable antelope; then past saber-horned roan antelope and zebras cantering over the rocky, rolling foothills of the scarp's apron; and on around an enormous monolith above which hawks and eagles soar. From there the Lufwashi remembers its way along more gentle slopes to its confluence with the Mwaleshi River.

Families of elephants standing in gallery forests lift their trunks, sniffing the air as we pass overhead; and thousands of buffalo pour from the woodlands into the shallow river to cool themselves and to drink. Rust-colored puku antelope, the size of white-tailed deer, are sprinkled across every sandbar, along with impalas, eland, hartebeests, warthogs, and every bird known to Africa, it seems.

When we reach the broad Luangwa, we see herds of hippos crowded bank to bank, blowing plumes of spray, their jaws agape at the airplane. Fat crocodiles, wider than a kitchen table, slither off the sandbars into the water. And not a sign of human beings.

Along the Mwaleshi again we find the poachers' track and follow its scribblings across the scarp's apron into the foothills and mountains. Delia notes the times, compass bearings, and topographical features that we will use to navigate back into the valley on the ground.

After we land back at the Mpika airstrip, Delia grins at me and holds up both thumbs. Never before have we seen so much wildlife in one place. Now we have to find out if it will be possible to live and work in this remote, rugged wilderness. We taxi to the side of the airstrip and tie down Foxtrot Zulu Sierra. For a few Zambian kwachas Arius, a toothless old tribesman and the government's keeper of the airfield, agrees to keep watch over our plane while we drive into the valley.

Following Delia's notes, we drive up the Great North Road until we find the rutted clay track that we hope will lead us into the park. It follows the northern base of the Kalenga Mashitu, a twenty-mile ridge of rock, through a cool, deep forest of spreading *Brachystegia* and *Julbernardia* trees. With their splayed limbs and luxuriant crowns, these trees dominate the classic miombo wood-land found at higher elevations throughout central Africa. Occasionally we see neat thatched huts nestled in the hills below Mashitu's rocky spine. Bending over each hut is an enormous green and yellow banana palm, providing shade, shelter from the torrential seasonal rains, and fruit for the family below it. From a settlement of about a dozen round huts, a gnarled old woman hobbles to the track holding up a bunch of bananas. As I am paying her for them, about thirty women and children gather behind her and begin to sing, their voices like wind chimes on the cool, moist air. After listening to three or four songs, we applaud the choir and drive on — while they are applauding us.

Soon after, the track forks and we stop to study Delia's notes. As we stand outside the truck, a small band of men approaches, the narrow blades of their hand axes hooked over their shoulders. The men curtsy with their hands clasped, eyes downcast in the traditional sign of respect, as we ask which track to take to Mukungule Village. Before answering, the spokesman grows an inch or two, then declares: "This track, she is good!" He hurries over to pat the ground of the left fork with both hands. "If you take it, you shall touch Mukungule." He smiles hugely, exposing his brown and broken teeth. Still hunched over, he rushes to stand on the other fork. "Aaahh, but this one, she has expired." His expression falls as he stomps on the expired track, as if to be sure "she" is dead.

"Natotela sana — thank you very much." We offer the only Bemba words we have learned, then drive away, leaving the men clapping and curtsying in farewell.

We take the living track and, four hours after leaving Mpika, it begins to wind through fields of maize and millet. We creep across a bridge of limbs and branches that snap, crack, and groan under the two-ton Cruiser, which sways drunkenly and threatens to break through to the water below.

Minutes later we "touch" Mukungule, its huts of ragged thatch and mud-wattled walls standing among maize patches overgrown with tall weeds and grass. The track leads us right past the fire circle of a family's boma, and even though we leave tire tracks through their "living room," they step back, laughing, waving, and cheering. "Mapalanye! Mapalanye!" The hellos of the women and the children's squeals of laughter mingle with the flutter and squawks of retreating chickens to create a raucous, but somehow musical, welcome. Several women, wrapped in brightly colored chitengis, pause from "stamping their mealies," their long poles poised above the hollow tree stumps they use as stamping blocks, or mortars, for crushing the maize kernels. An older woman sits on a stump in front of her hut, her foot working the treadle of an old Singer sewing machine as she stitches a brightly patterned cloth.

A throng of young people crowd around our truck as we stop. One lad softly and shyly strums his guitar, homemade from a gallon oil can, with a rough-hewn wooden neck and crude wooden tuning pegs. Nails driven into the neck and bent over under the wire strings form the frets. With a little encouragement from us and his friends, he begins a twangy tune. We listen intently for a while until it begins to seem that this song has no end; we slip away.

As we pass Munkungule's last hut, the grass in the track is suddenly taller than the truck. I stop, and Delia climbs up to ride on top so that she can guide me. An hour and a half later, but little more than six miles farther, the track forks again. Ahead of us on the left the four mud-wattle and thatch houses of the Mano Game Guard Camp pop up like mushrooms growing out of the tall grass and maize patches.

Set on a barren acre above the Mwaleshi River, this camp is home to four game scouts and their families. Four hundred yards from the main camp, at the base of a small kopje, are two other houses and a storeroom for the "Camp-in-Charge," his deputy, and their families. In Zambia game guards, or scouts, are civil servants who are given military-style training in firearm tactics, wildlife law, and a smattering of ecology, then charged with patrolling the country's national parks and other wildlife management areas to guard against poachers. Gilson Kaweche had told us that there are four other scout camps, spaced about twelve miles apart along the western boundary of the park; but together they have only seven scouts. Mano, with its six scouts at the center of the chain of camps, is the only one with enough men even to mount patrols. In fact, the Lufishi camp has been closed down, its single scout suspended for collaborating with commercial poachers. In all, thirteen scouts are charged with protecting the North Park — an area larger than Delaware.

We take the left fork and I stop the truck near a circle of twelve to fifteen men sitting on the bare red earth of the main camp. They look up at us with somber faces, their eyes red and watery. In their midst a large clay pot is brimming over with the frothy local beer; several reed straws stand in the mash. One of the men is wearing a pair of green uniform trousers, suggesting that he is a game guard; the others are dressed in tattered shirts and pants, probably obtained from local missionaries. After greeting them I ask for the Camp-in-Charge, and a stocky Zambian with prominent ears and black hair graying at his temples slowly stands up and walks unsteadily toward us.

"I am Island Zulu, Camp-in-Charge," he announces grandly, his head cocked to one side, as I hand him our letter of introduction from the director of National Parks. A man with a red bandana wrapped around his head saunters up.

"I am Nelson Mumba, Camp-in-Charge at Mwansa Mabemba," he says through a crooked smile, one front tooth missing. "We have no food or ammunition for patrolling, and no transport. We are supposed to be given mealie-meal every month, but it never comes," he complains. "Our families are hungry. Even now our wives are working in the fields so that we can eat." With his bandana, he looks like a pirate as he points to a group of women hoeing in a nearby maize patch.

"That's terrible!" Delia commiserates. "Have you told the warden?"

"Ha! The warden," Zulu shoots back. "He cares nothing about us. He has not been here in more than two years."

At this point I'm not sure what the scouts expect us to do about their problems. I explain that we are looking for a site for a major project, and if North Luangwa turns out to be the right place, we will help them all we can. Mumba mutters something, spits into the dust, and they all walk back to their beer circle. As we pull away in our truck, they are sitting down at the beer pot, reaching for the straws.

After fording the clear, rushing waters of the Mwaleshi River, we camp near a small waterfall hidden in the deep miombo woodlands. From here the Mwaleshi tumbles over the three-thousand-foot scarp mountains, and we will have to do the same. The thick forests prohibit us from following the river, so we will have to find it again when we reach the valley floor.

To test whether or not we can work in North Luangwa, we will try to drive down the scarp, then along the Mwaleshi to the Luangwa River and back. From our reconnaissance flight, the floodplains along these two rivers appear to be among the most important habitats in the park. If we cannot get to them, there is probably little reason to settle in North Luangwa. This trek will not be easy, for most of the way there is no track.

No one who cares about us knows where we are going, or for how long. Our Land Cruiser is nearly worn-out; we don't have a radio, a firearm, or fresh antivenom — none of which are available in Zambia even if we could afford them. In an emergency, it will be a minimum twenty-hour drive to the nearest hospital in Lusaka

— which is often critically short of everything, including AIDS-free blood. Despite all of this, we decide that having come this far we may as well go ahead.

Early the next morning we snatch our mosquito net from the tree limb above our bed on top of the truck, stuff down some raw oatmeal, and start driving. The trail is gentle at first, wending its way through the lush miombo (*Brachystegia*) forest with tropical birds flitting overhead.

But as we round a rocky outcropping, the track abruptly disappears; we will have to drive over the side of the mountain without one. A steep slope, studded with jagged rocks and deep ruts, drops off through the woodlands in front of us. Immediately the truck charges forward, going too fast. I slap it into low gear, but the heavy trailer lurches forward, ramming the Land Cruiser in the rear. Its back wheels heave off the ground, sliding sideways into a jackknife. The drums filled with aviation fuel slide forward, slamming against the trailer's front gate. Spinning the steering wheel, I gun the engine to keep the truck ahead of the trailer. The Toyota sways heavily, rumbling faster and faster over the boulders as I pump the brakes on and off. Still wet from the river crossing, they are not slowing us.

"Get ready to jump!" I shout to Delia. She grabs for her door handle as we rattle and bounce down the steep grade.

I stand hard on the pedal until the brakes begin to hold. Fighting for grip, the tires clutch at the sharp rocks embedded in the slope. Thumb-sized chunks of tread tear loose with a popping sound.

We bottom out of the quarter-mile grade bouncing and barely under control. We are going much too fast. But every time I jab at the brake pedal, the rig tries to jackknife. Finally, as I desperately feed in just enough brake to slow us down, but not too quickly, the truck's rear wheels settle back onto the slope and stay there. Shaking her head, Delia relaxes her grip on the dash and I release my choke hold on the steering wheel. The Luangwa has taught me my first lesson: get into low and go slow-slow when descending the Muchinga Escarpment.

Over the next hour and a half we descend three more steep

pitches and many smaller ones, until it feels as though the truck is standing on its nose. Finally we drive out of the shadows of the miombo woodlands onto a rocky ridge with a panoramic view of the valley: miles of golden grassland cover the rolling knolls of the scarp's apron and the valley floor in front of us; the mountains of the Muchinga Escarpment curve away to our right, disappearing in the distance. Chinchendu Hill, a giant two-by-four-mile monolith eight hundred feet high, juts from the valley floor about five miles away. In the language of the Bisa tribe, "Chinchendu" refers to a big man who stands firm, broad, and tall. To our left, about six miles away, a conical hill resembling a rhinoceros horn is shrouded in blue haze from the heat and smoke of wildfires sweeping the valley. Locally known as Mvumvwe Hill, it and Chinchendu will be our two main landmarks as we explore this part of the valley.

Not many tsetse flies bothered us on our drive down the scarp. But by now we have dropped nearly two thousand feet, and the temperature has risen at least ten degrees to about 87°F. Tsetses swarm inside the Cruiser, biting every exposed patch of skin, even through our shirts, shorts, and socks. Delia soon counts twenty-one bites on her legs. Too hot to roll up the windows, we beat at the flies with our hats, crush them against the windshield, and finally light cigarettes Delia brought for bartering with soldiers at roadblocks along the main road from Lusaka. We puff like fiends until the blue nicotine smog forces the flies to retreat.

The truck and trailer jolting and clattering, we drive on toward the confluence of the Mwaleshi River and the Lubonga, its smaller tributary. Two hours after leaving the scarp, we stop to study Delia's notes again, but cannot determine which way to go.

Walls of *Combretum obovatum*, a thorny scrub that stands twelve feet high, stretch across our path. Twisting and turning along the dusty valley floor, we try to navigate through the maze of thick brambles. Time and time again we fight our way through a brier patch only to find a deep, ragged stream cut blocking our way. Standing on the steep banks, looking down at the uprooted trees lying in these dry washes, I remember the flash-flood warnings Norman Carr gave us in Lusaka: "Don't get caught in the valley after the rains come in November. If you do, you might not get out — until it dries up in May or June."

Large and small hoof prints cover the ground, but we see few animals. All the vegetation except the crowns of the trees is parched dun-brown by the sun and heat. Leave it to us to discover another desert!

Nearly three hours after leaving the scarp, the soothing blues and greens of two rivers, one on either side of us, wink enticingly through the tangle of dry scrub. A bit later we push through a stand of tall grass and on our right is the Mwaleshi, its white, sandy bottom showing through sparkling water; on our left is the Lubonga, its tributary, little more than a dry-season trickle. We have arrived at the precise confluence of the two rivers.

A small herd of puku — freckles of red and brown in the brilliant green grasses — stand on the riverbank forty yards downstream; and fish eagles sit high in the treetops along the wide, shallow river. After the desolation of the obovatum scrubland, we drink in this scene, as we will the water; and the squints and frowns we wore from the glare only moments before dissolve from our faces. We run to the Mwaleshi, scoop it up in our hands, and douse our faces and necks. But it is not enough. Leaving the heat behind, we jump off the bank, fully clothed, into the water. Crocs be damned.

The sun is setting by the time we cross the river. High on a bank above the confluence, we unroll our mattress on the top of the truck, set out our table and two chairs, and light a campfire. It has been rough going, but we've made it halfway to our objective: the Luangwa River. To celebrate I splash some "pirate" rum, made in Zambia, into the orange squash in our plastic mugs, and we toast our first night in the valley. Below us herds of puku and zebras drink from the river, its water the color of molten steel in the glow after sunset. A neighborly kingfisher hovers at eye level, then tucks its wings and arrows into the pool at our feet. Later, after a dinner of canned chicken stew and a bottle of Drankenstein wine, we climb into our sleeping bags on top of the Cruiser.

Each night for weeks Delia and I have been playing a game, and tonight is no different. Burrowing down in my bag, I sigh, "Boo, tonight I'm going to show you a shooting star, or at least a satellite. This is it — get ready — I know we're going to see at least one." We gaze up at the heavens, looking for a sudden streak of blue or a faint yellow spot of light moving faster than the more distant stars. Within a few minutes we are asleep, somehow saddened that we have seen neither. In the Kalahari, where the arid skies are much clearer, it would have been different.

During the next two days we search up and down the Mwaleshi for a temporary campsite and landing strip. We find only one place where the loops in the river course are far enough apart to allow room for takeoffs — especially those lengthened by the heat and the resistance of rough ground and grass. The place we choose would make Cessna's insurance company shudder. It is a comparatively level surface, but cut short by the river at one end and a tree-covered hill at the other. Worse yet, a big sausage tree, thirty feet high, stands in front of the hill near the end of the strip.

I pace the runway several times, but cannot stretch it to more than 338 yards. According to the pilot's operating handbook for the Cessna 180K, a takeoff on a grassy surface in this temperature should require 330 yards; then add another 215 yards before the plane can clear a 50-foot obstacle — a tree, for instance. If we're hot and heavy, we may have to fly around that tree after takeoff. If not, well, we've got 8 yards of runway to play with and another 200 or so to clear the tree.

Using axes, shovels, and picks, we spend a blistering hot afternoon chopping out tough, spindly shrubs and leveling off termite hummocks to make our airstrip. To finish it, we mark both ends of Airstrip One with piles of buffalo dung. Before dawn the next morning we unhitch the trailer, stash it in tall grass near the airstrip, and head for Mpika to get our plane. Without the trailer to get us stuck in every sandy crossing, the drive back up the scarp to the village takes only eight hours. Just thirty minutes after takeoff I land on our new runway. Because the drive back will take Delia much longer than my flight, she will have to camp along the track and meet me here tomorrow.

I drag some thornbushes close around Zulu Sierra's belly, then spread my bedroll on the ground under the plane. A few years ago

Norman Carr's best friend was pulled out of his tent and eaten by a hungry lioness in the Luangwa Valley. So I am a little wary of sleeping in the open. In the Kalahari we often slept not only on the open savanna, but *with* lions. In this unfamiliar habitat, however, my primate ancestors, speaking to me down long lines of evolution, are warning me to be careful.

In the late afternoon I sit on the riverbank, watching a pied kingfisher dive for dinner. Deception Valley seems of another world, another time. It hurts to remember how much we left behind there, in the Kalahari: Sunrise, Happy, Stormy, Sage, and the other lions; and Dusty, Pepper, Patches, and Pippin, our brown hyenas. What can ever replace them, or fill the need they have created in us?

Night shadows begin stalking the riverbanks. A Goliath heron drifts by, its wings whispering in the still dark air. Through the quiet current a large V-shaped ripple cruises slowly upstream toward me. I retreat to my plane — my tin of technology — and sit by my campfire, out from its wing. A lion calls from upriver, another answers from down. Then it seems too quiet, so I lie under the plane, my three-foot-high thornbush boma pulled around like a comforter. Soon I am asleep. And I dream of another land, with larger lions, with deeper roars; of standing with my arm across the shoulder of a big male named Muffin. Together with Delia I look to the distant horizon, far across the dunes.

o o o

"Munch — munch — munch." Roused from sleep, I slowly raise my head and look out from under the plane. Four hundred fifty buffalo are mowing and fertilizing the gravel bar along the Mwaleshi River where I landed last evening. They are headed straight for me, some of them only fifty yards away. I crawl forward and stand up, leaning back against the propeller and peering over my thornbush boma. None of the buffs notice me. Their broad muzzles pressed to the ground, tails flicking while ox peckers flit about their backs, the mean machines mow on. Now and then one snorts or grunts loudly, shaking its bulldozer boss and wide-sweeping horns to shoo the flies, slinging saliva onto the grass.

The airplane means nothing to these buffalo since they have never seen one before, and because I am frozen against the propeller they haven't yet distinguished my human form. Buffalo have a hard time seeing anything that isn't moving, even at close range, and I am downwind from them so they haven't smelled me.

Grumbling and mooing, they continue to graze toward me. By now the closest cows are only twenty yards away, and I can smell their dank, musky scent. People who surprise buffalo at close range may get gored, stomped, tossed, and even chewed. Last year two game scouts from Mano were killed by them. At ten yards the nearest buffalo are too close, but they still haven't noticed me — and they are still coming. The only thing to do is to stay still and hope they move on.

But tsetse and dung flies are crawling over my face, and the urge to brush them away is unbearable. Cautiously I begin raising my right hand toward my chin. The lead cow, not more than eight yards away, lifts her head, stops chewing, and looks directly at me, a large sward of grass jutting from the corner of her mouth. The wrinkles over her eyes deepen, her muscles stiffen, and air explodes from her black nostrils. Some large bulls at the edges of the herd immediately lift their heads and key on her. I stop my hand midway up my chest.

Fifty buffalo are now staring at me. Dung flies are crawling over my cheeks and forehead, sucking moisture from my nostrils and the corners of my eyes and mouth. Tsetse flies are biting my neck and arms. I don't move.

The cow in front of me relaxes and lowers her head to the grass, but the older, more experienced bulls behind her stalk forward, sighting down their black cannon-sized muzzles at me. Close to the cow they stop, still glaring at me. My hand crawls to my chin, covers my nose and mouth, and freezes there. The bulls snort loudly, shaking their heads and stamping their feet.

I allow my fingers to wander over my nose and cheek, chasing away the maddening flies. The bulls snort and stomp again. The cow raises her head. She shakes her horns and my stomach tightens.

Then she spins away and clomps off, stopping fifty feet away to look back at me. But the bulls still move toward me, snorting and stamping the ground with their heavy hooves. I wipe my face with my hand and then twiddle my fingers at them. They stop again, raise their heads, then lower them, never taking their eyes off me. The cow begins to graze. Two of the males swing their heads to look at her, then they also graze.

I too relax. Moving slowly, I poke up the fire and minutes later pour a pan of boiling water over some coffee grounds. Then I sit down under the nose of the airplane, savoring my hot, thick "camp coffee" as I watch the herd graze past me. Some are so close I swear I can see the plane's reflection in their eyes, see puffs of dust around their nostrils, hear the coarse grass tearing as their teeth crop it off.

The sun rises slowly behind the herd, cradled between the banks of the river, setting fire to the fringe of hair around their ears and the whiskers on their muzzles. My coffee tastes especially good this morning, in North Luangwa.

Late afternoon. I am still wound up from the morning's encounter. Delia arrives from Mpika in a cloud of dust, and as I tell her about my communion with the buffs, she rolls her eyes and sticks her tongue in her cheek. Fortunately there is plenty of "B.S." to prove my story.

Up at the first hint of dawn, we load the trailer with the gear we won't need on our reconnaissance to the Luangwa, and stash it in an obovatum thicket. We enclose the plane in a thornbrush boma again — to keep hyenas and lions from chewing its tires and tail — and by seven o'clock we are on our way southeast along the river. We have traveled less than a mile when a mean range of scrambled hills blocks our way. They are covered with brambles and chopped up by dry stream channels; there is no way around them. The Mwaleshi is too wide and full of quicksand for us to ford it. We will have to drive over these "Hills of the Chankly Bore," as we call them.

Near the river a narrow gully leads up a steep slope and through the leading edge of the hills. Here we ax down hummocks and chop out briers. But when I try to climb the grade with our rig, I lose traction halfway up. So I back down and try again — and again. Finally, we winch up the truck first, then turn it around to winch up the trailer.

Only minutes after we've descended the Hills of the Chankly Bore, a wide, sandy river lies across our path. No matter which way we turn, a gauntlet of steep slopes and stream and river cuts blocks our way. By placing short planks in front of the wheels to keep them from sinking, we finally make it across. Fifteen minutes later we are down to our axles in mud on the edge of a lagoon. Almost as soon as we have winched out, we drive a brittle mopane stump through one of the Cruiser's tires while trying to push through a solid wall of obovatum briers. We change the tire and hack our way through the thicket with machetes, but then the truck's right front wheel falls into a deep hole hidden in tall grass. At one point it takes four hours to go a thousand yards through a woodland of dead snags.

Through all of this the Mwaleshi flows serenely by our side, twinkling in the sunlight. It is as though the river is teasing us during our ordeal. We see an occasional puku or waterbuck, but where are the mighty herds of buffalo, like the one that surrounded the plane, and the zebras, eland, and impalas we saw from the air? Eventually ten eland trot by, watching us as if they were spectators at a one-truck demolition derby. But their magnificence is lost on us: we are down to our axles in sand again, and shoveling furiously in the 110°F heat. Each obstacle we cross we will have to recross on the way back. There will be no quick way out.

Every evening we park our truck under a tree with a limb at the right height for our mosquito net, climb to the Cruiser's rooftop carrier, take down our chairs and folding table, stand the high-lift jack and spare tire out of the way, and lay our sleeping bags on mats on top of the truck. I have already lost a pair of boots to hyenas, so we keep most of our gear inside the truck or on top of it while we sleep — surrounded by our boots, extra clothes, toothbrush kits, towels, jacks, and spades. Stacked on the front seat are binoculars, books, and camera gear to keep them from getting soaked by the dew. Each morning everything has to be put away before we can move on. We have been living like gypsies for four months, exploring wildlife areas in Zimbabwe and Zambia, and we are anxious to settle somewhere.

On the fourth morning we drive through a dense woodland. As we emerge on the far side, a forest of stark, dead mopane trees and utterly bare ground stretches before us. In this "Torrible Zone," as Delia calls it, the trunks and branches of these skeletons — peeling, split, and rotting — all seem to have been killed at the same time, frozen in the act of life by some cataclysm. The gnarled roots clutch at soil baked by the sun to a dusty pancake. As far as we can see, not a single blade or leaf of green promises life.

"Mark, look over there!" I stop the truck and we walk to a thin grove of trees near the edge of the dead woodland. Five elephant skulls, bone white and half the size of bathtubs, are scattered about the area with pelvises, leg bones, ribs, shoulder blades, and other remains. Horrified, we notice skeletons lying everywhere: one here, five over there, six there.

"The bastards!" I kick the dust.

Hurrying from one skull to the next, we examine each one. All have small holes, where small holes should not be in elephant skulls. All have had their faces chopped off, their tusks hacked away.

Now we understand why we have not seen a single living elephant, or a sign of one, in the eight days since we entered the park. We are standing in the midst of a killing field, where gangs of poachers have slaughtered every great gray beast they saw. It is an elephant's Auschwitz. Looking around at the carnage, I can't help but wonder if the death of these elephants might somehow be related to death of the forest. The magnitude of the poaching problem in North Luangwa hits us like a fist in the stomach. Although we have not yet run into poachers, it must be only a matter of time until we do. There will be no ignoring them, running from them, pretending they do not exist. If we stay here to work, we will have to do something about them.

Sobered, we continue to follow the Mwaleshi River toward its confluence with the Luangwa. For hours the bush is so thick we cannot see the river, and we stay on course only by using a compass. We are driving through tall grass and obovatum thickets when suddenly the front of the truck falls out from under us with a crack. The Cruiser jams to a stop like a shying horse and pitches us against the windshield. We rebound into our seats and rub our heads, coughing in the haze of dust rising from the truck's floor.

After checking to see that Delia is not seriously hurt, I open my door, which is now at ground level, and get out to check the damage. The right front end is sitting bumper deep in a washout that was hidden in the grass. The spring has broken, letting the chassis down on the axle. If the axle had snapped, we would be stranded. We don't have enough spare blades to fix the spring properly; I'll have to jerry-rig something.

With the high-lift jack we raise the truck and block it up on pieces of dead wood. Using a small battery-powered drill, I make center-bolt holes in three tire levers, then wire them in place where the broken spring blades were.

Two hours later, the truck is ready to roll again. "I can't believe we still haven't reached the Luangwa," Delia sighs, slumping down in the grass, her arms covered with the claw marks of the obovatum, matted hair hanging low over her forehead. "Let's discuss this a bit. It's taken us four days so far. We have no radio, nobody knows where we are, we don't know where we are. If the truck breaks down and you can't fix it, it's going to be a long walk back to Mpika. We're digging ourselves in deeper and deeper."

I sit down beside her, pull a grass stem, and begin chewing on it. At last I say, "That's what we've been doing for years."

"Right; fine. I just thought we ought to stop and consider for a minute." A silent minute later — "Let's go," she says.

After another hour we finally break through a hedge of obovatum scrub into an avenue of trees growing on a high bank above the sandy Mwaleshi. Standing on the top of the bank, we are greeted by a sweeping view of the river valley. The sun is setting over the escarpment, its rays in a fiery dance, skipping over ripples and sandbars in the water. The floodplains near and far are spotted with wild animals: six hundred buffalo grazing across a grassy plain; fifty zebras ambling toward the river to drink; a herd of waterbuck lying on a sandbar downstream; impalas browsing at the edge of the mopane woodland. Nearby a herd of Cookson's wildebeest — found only in Luangwa — gallop about in a sun dance. "WHOOOO — HUH — HUH — HUH! MPOOOSH!" The sound, like a humpback whale playing a bassoon, echoes from our left. "Hippos!" I grab Delia's hand and we run downriver toward the calls. Not more than a hundred yards away, we round a last bunker of bushes and there is the Luangwa, with the Mwaleshi flowing into it. Where the two rivers join is a large pool crowded with a hundred hippos, their piggy eyes on us, their nostrils blowing plumes of water in the setting sun as they twiddle their ears.

After our tangle with the bramble and the broken woodland, Africa has won us back.

We sleep with the hippos that night, their grunts, sighs, hoots, and bellows the refrain of a strange orchestra in the river below. And in the forest near our bed, a leopard's hoarse cough sets off shrieks, screams, and chattering from a troop of baboons. Later, just after I have gone back to sleep, the cool air draining downriver from the mountains of the escarpment carries with it the trumpet of an elephant, and finally the heavy, insistent roar of a lion.

Born in Tanzania, the Luangwa sweeps between fifteen-foot banks and past broad sandbars for 460 miles southwest through Zambia. Like a snake, the river writhes, twists, and even coils back on itself in sharp hairpin turns, its water occasionally breaking through the land neck of a turn, or silting across the mouth of a sharp bend to pinch off an oxbow lagoon. A few tracks, but no major roads intrude on this stretch. It is one of the wildest rivers left in Africa.

During the next three days we make our second airstrip, Serendipity Strip, on a long floodplain near the Mwaleshi-Luangwa confluence. When it is finished, I fly the plane down from Airstrip One, build another bush boma around it, and then by truck and plane, Delia and I begin exploring north along the high banks of the Luangwa.

Every day we are more and more convinced that we want to live and work in North Luangwa. No one has ever done research on animal behavior or conservation here, and it seems a good place to test the concept that if the villagers nearest the park receive direct benefits from its wildlife, they will want to help conserve it. The people of the Bisa and Bemba tribes should be no exception. But every time I contemplate hauling tons of avgas and other supplies over the scarp, especially during the rains, trying to track research subjects through a maze of stream cuts, or persuading a dozen underequipped game guards to protect such a huge area from well-armed commercial poachers, I just shake my head. True, we have made it to the Luangwa; but we are not yet convinced that we can operate here.

We have carried so many spare parts, tools, and extra jerry cans of fuel with us that there was too little space left for food. Our supplies are already low, and we face a major trek back to Mpika for more. Three days after setting off north along the Luangwa, we chart a course back to Serendipity Strip. Two days later, in the late afternoon, we are lying in the hot, shallow water of the Mwaleshi at our first base camp, our toes and noses sticking above the water.

o o o

The sun is high and hot by early afternoon when we say our good-byes on the riverbank. I check my watch as Delia begins driving to Airstrip One. She will go halfway this afternoon, then camp for the night and meet me there when I land tomorrow. Sleeping again under the belly of the plane, I am both comforted and aroused by Africa's symphony of the night.

My only visitor is a honey badger, who snuffles around my feet as I sit on the plane's main wheel having my cup of early-morning coffee. In the afternoon I fly northwest along the Mwaleshi and spot our truck crawling along below, approaching the Hills of the Chankly Bore. Our timing has been almost perfect; Delia will arrive at Airstrip One only minutes after I land.

But she does not. Standing under the wing, shading my eyes against the late-afternoon sun, I can see our truck and trailer three-quarters of a mile away, entering the steep cut through the last hill before the airstrip. Then it disappears. Half an hour later Delia still has not arrived. Something is wrong.

I walk and then run toward the hills, dodging bushes and stumbling through the chest-high grass. Once Delia started down the cut, she could not have stopped the heavy rig on the slope's loose gravel; one way or another she must have made it to the bottom. But I cannot see the truck or trailer through the undergrowth — until I round the last thicket. The Cruiser is on its side, jammed between the high, narrow sides of the ravine, its roof downslope. Behind it the heavy trailer has jackknifed into the bank. Delia is nowhere to be seen.

"Delia! Are you all right?" I yell as I sprint toward the truck. The only sound is the hiss of battery acid spilling onto the hot engine. "Delia . . ." A stirring comes from inside the Cruiser.

"Oh no, Mark! Look what I've done! I just couldn't stop it," she moans, stepping out through the driver's window.

"Don't worry about the truck." I smile. "It's easier to service on its side." She forces a weak grin as I hug her.

She tells me that she stopped at the top of the steep ravine, shifted to first gear, and eased the Cruiser forward. But I had forgotten to warn her that I had unlocked the trailer's brake, to make it easier to maneuver in reverse; and as she hauled the heavy rig over the lip of the slope, it leaped forward, compressing its hitch and slamming into the rear of the truck. Jammed back in her seat by the impact, she pumped the brake and fought the steering wheel as the Toyota rolled ever faster into the divide. The walls of the cut were so close to the truck that she could not jump out without being crushed. Trapped inside, she tried to ride it out.

Halfway through the chute and still gaining speed, the truck's right front wheel struck a rock embedded in the wall of the ravine. The steering wheel tore itself from Delia's hands and spun sharply to the right. Immediately the truck turned and began to climb the steep canyon wall. At the same time the top-heavy Toyota rolled dangerously to its left. Delia grabbed the wheel and clung to it, trying to straighten the truck, but was not strong enough.

The trailer forced the truck farther up the wall, Delia still pulling frantically on the steering wheel. The Cruiser rolled onto its left side. Shoved by the trailer filled with avgas, it continued down the slope, its roof leading the way. Inside, Delia was thrown out of her seat, slammed against the roof, and battered by an avalanche of toolboxes, food crates, and camping gear. After the rig slid to a stop, she clawed her way out of the rubble. She was dazed, but fortunately not seriously hurt.

o o o

Using the little daylight left, we organize the jacks and winches that will set the truck back on its wheels, then sleep on the river-bank near the overturned vehicle. Before sunrise the next morning we rig the high-lift jack under the rooftop carrier. I run the truck's winch cable through a pulley that I've secured around a tree at the top of the hill. With poles cut from nearby trees, we fashion a sling for the Toyota's canopy and attach to it a yoke of chains. Then we hook the cable to the yoke.

Before trying to lift the truck, we tie the trailer to another tree upslope from the rig, so that when the Cruiser finds its wheels it won't run away again. Alternately jacking and taking up on the winch, we slowly lift it. About noon it staggers back upright. We fill its battery with water from the river, add some oil to the trans-mission, and crank it up. Aside from minor damage to the left front fender and the canopy, it is fine. Delia comes out even better, with only a few bruises and scrapes on her left arm and leg.

Nevertheless, turning over the truck seems to solidify all the doubts about working in North Luangwa that have been simmer-ing in our heads for days. Granted, the park is beautiful and it has an incredible variety of wildlife. But we can no longer shake off the thought that, true to its reputation, North Luangwa is too rugged, too remote, too inaccessible. That we are tired and really need to find a place, and that there are few comparable wilder-nesses left in Africa, does not make the rivers easier to cross or the slopes easier to climb. To drive even a short distance we have to make war on the landscape, mount a major expedition. And this is the dry season. Everyone has told us that when the rains come, we will be flooded out. It is a hard day's trip to Mpika for basic supplies, and aviation fuel will have to be brought from Lusaka, sixteen hours away. The poachers shooting elephants in the park may not take kindly to our plans, and because our licenses have not been approved, we have no firearms. On our limited budget, the nightmare logistics seem almost insurmountable.

The bruised truck standing to one side, we share a can of beans for lunch while sitting in the warm water of the Mwaleshi. Our conversation keeps turning to the hope that our permits for Tanzania will have arrived in Mpika. "If not, I think we should just get our things, pull out, and head for Tanzania as tourists," Delia echoes my thoughts.

After repacking the truck, we camp at the base of the Hills of the Chankly Bore. The next morning I awaken to the sound of splashing in the river. Lifting myself on one elbow, I see two big-maned lions thirty yards away, romping across the Mwaleshi, kicking up spray as they whirl and slap at each other. Their powerful bodies reflect the new light of the morning sun and at least for the moment make us forget the rigors of the previous weeks.

Not more than an hour later, on our way to Airstrip One along one of the Mwaleshi's floodplains, Delia puts her hand on my arm and points to a forked tree leaning out from a steep gravel bank. Four lion cubs are tumbling in play around the base of the large *Trichelia emetica* tree. As our truck creeps closer, they climb a short way up the two trunks of the tree and peer curiously at us with bright, round eyes. Below, the ears and eyes of three lionesses slowly rise above the tall grasses.

When we drive closer, the cubs climb down from their perches. One of the females lowers her ears, turns, and disappears; another lays her ears back slightly and looks away, as if mildly annoyed and determined to ignore us. The ears of the third stay fully erect, and she raises her head a bit higher. Twenty yards from the tree I switch off the truck and we sit quietly, letting the lions get used to us.

All at once the cubs come to the edge of the grassy patch and peer at our truck. First one, then the others, come slowly toward us on stumpy legs, their eyes amber-colored pools. When the first cub is ten feet away, he stops and stares through my window, raising and lowering his head as though trying to get a better look at me. Each cub smells the right front tire and circles the Cruiser, eyeing it up and down. Then they waddle back to sit between the forelegs of the lionesses in the grass.

A little later one of the lionesses leaves the others, climbs a low termite mound, and sits watching for prey. We drive over and park just twelve yards from her. We name her Serendipity. It has been fourteen months since we were last close to lions, and it is as though some drought has broken.

I look at Delia and smile. "How many places are left in Africa where we can wake up with four hundred buffalo around our bed, golden lions romping through the river near our camp, and a lioness sitting beside our truck?"

Although we are down to our last tin of beans and have barely enough diesel to get us up the scarp, we decide to stay one more night in North Luangwa before going on to Mpika. Perhaps we will see Serendipity's pride hunt later this afternoon. We set up camp on the banks overlooking the confluence of the Lubonga and Mwaleshi rivers, in the same spot where we slept on our first night in North Luangwa.

When the heat finally breaks late in the afternoon, we climb into the Cruiser to search for the lions. "Elephants!" Delia points to where the river curves eastward. A small family of six elephants is walking out of the forest, heading for the river five hundred yards downstream from us. The largest female has only one tusk. They stop, lift their trunks to test the air for danger, then take several more steps before stopping again, one foot raised, their trunks swiveling like periscopes. They are nearing the water's edge when suddenly One Tusk whirls in our direction. Flapping her ears wildly and swinging her huge bulk around, she runs back into the woodland. The others follow and within seconds they have disappeared.

The river must have carried our scent to them. We are barely larger than dots on their horizon, yet they vanished as soon as they sensed us. Constantly harassed by poachers, they are so frightened by humans that they will not drink even at such a distance.

At this moment, in August 1986, we pledge to each other: no matter what it takes, or how long, we will stay in North Luangwa until the elephants come to drink at the river in peace.

But to stay we must find a way to survive the floods.

6

Floods

DELIA

> Except by the measure of wildness we shall never really
> know the nature of a place.
> — PAUL GRUCHOW

○ ○ ○

SPEARS OF SUNLIGHT stab through the forests of the Muchinga
Escarpment as I steer the old Cruiser carefully down the rutted
track into the Luangwa Valley. Riding with me is Chomba Simbeye, a wiry, twenty-one-year-old Bemba tribesman who knows
this part of the valley well. Mark will fly to meet us in three days,
assuming Simbeye and I can clear Airstrip One for a safe landing.

Instead of being gone for three or four weeks as Mark and I had
imagined, it has taken us more than a year to obtain all the permits
to operate in North Luangwa. Now it is late October 1987, leaving
us only a month to build an all-weather airstrip and base camp
before the rains and floods come. Unless we finish the strip by
then, we could be stranded for months.

We have been told that the poachers operate on foot and are
especially active during the rains. That's another reason why we
are determined to get settled in the park before then. We're not
sure how we'll stop the poaching, but we've got to find a way.

At the Mano Game Guard Post a few scouts laze about, while
their wives suckle infants, wash clothes in brightly painted basins,
or pound maize for the evening meal. With a slab of whittled wood,
one of the women stirs her simmering stew in a battered tin pot,
occasionally flavoring it with a sprinkle of blossoms from a large
straw basket.

Because I am unarmed, Mark has insisted that I take a game
guard with me into the park. Tapa, a tall, slender man with shy,

round eyes, volunteers to come. While he prepares his *katundu*, or belongings, the other scouts complain to me that they are still not getting ammunition or food from their command headquarters in Mpika. Just hours earlier the warden, Mosi Salama, told me that Mano had been given its monthly rations. I don't mention that we've heard from an honorary ranger that scouts often sell their government-issued food and ammo, or use the weapons to poach. If we are going to stop the poaching in the park, we will have to win the scouts' cooperation. Like anyone else, they need encouragement, equipment, and identity before they will do their jobs. If the government can't afford to give them food, medicine, and decent housing, we will raise the money somehow. Then they will no longer need to poach and will begin patrolling. My daydreams ease me down the rugged scarp.

After three hours of driving the familiar steep, rocky grades the three of us reach the confluence of the Mwaleshi and the Lubonga, that enchanted place where Mark and I vowed to stay in North Luangwa. A herd of puku lie on the sandy beach and a few waterbuck stand ankle deep in the slow-moving current. I would love to linger a while, but we are still two miles from Airstrip One and I want to get there by sunset. The dry season has reduced the Lubonga to a trickle of clear water flowing over ripples of loose sand. The old truck bores through, but the trailer bogs down in the wet sand midstream. We unhitch, and abandon it until later.

Just as the sun melts into the purple mountain peaks, we round a small stand of forest. The wide, shallow waters of the Mwaleshi sweep past tall banks on the far side, and the floodplain on our side opens into a sheltered grassland, tucked between high buffs. Several piles of bleached bones lie half buried in the tall grass — all that is left of Airstrip One.

There isn't enough time before dark to set up a proper camp; we'll just prepare for the night. Tapa and Simbeye gather wood, build the fire under a large fig tree, and set up the chairs and tables. I unpack my bedroll, put water on to boil, then walk the fifty paces to the river to bathe. There is a steep bank between camp and the river, so I have my privacy. I undress, check carefully for crocodiles, and plunge into the clear, shallow water. As I roll along the sandy bottom, the heat of the day and the frustrations of the previous months float away, drifting down the Mwaleshi to the Luangwa, into the Zambezi, and on to the Indian Ocean. I feel free, alone, strong, and happy. I laugh out loud as I splash.

Dusk is deeply upon us when I climb back up the bank, and Simbeye warns me that it is very dangerous to stay in the river so late. I thank him for his concern and lie that I will be more careful in the future. He and Tapa have built such an enormous fire — to frighten away lions, they explain — that it lights up the entire canopy of the fig tree and makes the already warm camp unbearably hot. The heat drives us to the edge of camp — much closer to the lions, if any are about — where they teach me words in their language until dark.

Simbeye, a cheerful, self-confident young man, is from the village of Shiwa N'gandu, which means Lake of the Royal Crocodiles. In the past the chiefs of the Bemba met at the lake each year for a ritual croc hunt. Simbeye squats by our fire, boiling mealie-meal in a crusted, sooty pot to make n'shima, the staple food of his people. We eat with our fingers, dipping thick pasty balls of n'shima into a relish of beans and onions. Simbeye, speaking in a low, raspy voice, tells me tales of Bemba folklore.

Sometime long ago, he says, there was a mighty tribe of warriors in the land that is now Zaire. One day a strange woman, who had ears almost as large as an elephant, wandered into the village of the chief, Chiti-Mukulu. Most thought she was ugly, but knowing that she would make a good wife, Chiti-Mukulu married her. She bore him three sons, who indeed were very strong but were always getting into mischief and causing much trouble to the chieftain. When they became young men, Chiti-Mukulu banished them from his lands and they were forced to travel far away to the south. Some of the other warriors joined them and they started a new tribe, calling themselves the Bemba. After many months of plundering the villages of other tribes, they came upon a beautiful valley bordered by mountains, where wild animals thrived in thick forests. They chased out the local residents and formed Bembaland, which is now Zambia's Northern Province.

To this day the "paramount," or highest, chief of the Bemba

bears the name Chiti-Mukulu. It is never the son of the chief who inherits the throne, but his nephew. When I ask why, Simbeye explains to me that a man can never, never be certain that the son of his wife is his own child. But he *can* be sure that his sister's son is his true blood relative. It is one of the clearest cases of kin selection — passing one's genes to the next generation through kin other than offspring — I have ever heard.

Soon after we eat, I drive the truck some distance from the tree; I'm not going to miss the night sky because of the bonfire. Instead of making my bed on top of the Cruiser, I lay my safari mattress and sleeping bag on the grass. Tonight I want to sleep on the ground, between the earth and the moon. The truck will be close enough for a retreat if lions come.

As I lie next to the earth, the moon covers me with its platinum blanket, not bringing warmth like the sun, but a caress of hope. The silky light transforms every leaf and blade of grass on the plain to shimmering silver. Puku and impalas standing next to the river fade into subtle impressions against the pale sky. I smile before I sleep. I too am in my place, for I am a moon person.

Awake at 5:25 A.M., I watch first the hills and then the river awaken from darkness, stretching out in the orange-pink dawn. Just after six, Simbeye whispers, "Madam, wild dogs — at the river." I look downstream to see three wild dogs, their bold black, white, and brown coats standing out vividly in the sunrise. They gallop through the shallow river, splashing a spray of water against the sun.

By six-thirty we have packed our beds, eaten breakfast, and walked to the airstrip, a mere hundred yards away. We slash grass, fill in holes, level ant mounds, chop down small bushes, and clear away logs and stones. By eight o'clock the heat is unbearable: Every fifteen minutes I walk to the river, and wet my shirt and hat to keep cool. I have never liked Airstrip One; it is only half the length required by regulations. But I can't bear to cut down the enormous sausage tree at the southern end; Mark will have to avoid it, as he did last year.

At noon we break for lunch. In the midday heat waves, the far riverbanks wiggle and dance. While Tapa and Simbeye rest and eat in camp under the fig tree, I take a can of fruit cocktail, some sweaty cheese, the binoculars, and a book to the river. The shallow water is hot, but as soon as I am wet the breeze brings goose bumps to my bare skin. As I sit in the river reading my book and eating lunch, a pair of Egyptian geese land nearby and paddle around next to a sandbar.

More work through the afternoon: we clear the strip, mark the ends with new piles of bleached bones and buffalo dung, and cut thorn branches to make a hyena-proof boma for the plane. At dusk Simbeye and Tapa build another big fire, but again I camp far away, closer to the moon and the stars.

In the morning we complete the thorn boma, rescue the trailer from the sand, pitch the tent, and fill it with our tin supply trunks. Mark and I will have to continue sleeping on the truck or ground, since we don't have another tent. Using branches from a fallen tree, we construct little tables for the pots and pans, and bury the film cooler.

The same pair of geese keep me company at lunch — a good thing, because the heat makes my book boring. The work is finished, more or less. If it wasn't so hot I might do more. Instead I read until four o'clock, then explore the river to the south on foot. Simbeye and Tapa insist on escorting me, so we set off across the plain together. I stop abruptly.

"Listen. Moneni ndeke!" I shout. They echo me, "Moneni ndeke. The plane is coming."

We run all the way back to camp and I grab a roll of toilet paper to use as a wind sock. Mark circles overhead, waggling his wings in salute as I stand on top of the Cruiser, letting tissue play out into the slight breeze. Mark buzzes the strip once to check our work, then glides in for a perfect landing.

Simbeye, Tapa, and I show Mark around the little camp with more enthusiasm than it deserves. While the Bembas go to collect more firewood, Mark and I walk to the river to bathe. Together we frolic in the water, laughing and talking endlessly about our separate trips into the valley — one by air, one by road. Tomorrow we will search for a permanent campsite; we are not sure we could drive to this one during the rains. The pair of geese forage about in their spot by the sandbar, and I imagine that they are pleased that I too have a mate.

o o o

Wiping sweat from our brows, we stand on a high bank of the Lubonga, studying an old floodplain that runs for a thousand yards along the river. Fifteen huge marula and sausage trees shade the sandy ground. A dry oxbow, where the river once ran, surrounds the plain on three sides, and on the east bank the Lubonga trickles gently over the sand. But this is the dry season. If the oxbow and river flood when the rains come, the plain will become an island and could easily be cut off. Our twenty-year-old map definitely shows this stretch of plain as an island, with water on all sides.

For ten days we have searched from the air and hiked through the heat to locate a suitable base camp. The site must be accessible year round and have drinking water, shade, and a nearby site for an all-weather airstrip. This floodplain has water and shade, but the only possible airstrip is three miles away. And the question is whether or not it will flood during the rainy season.

"We're looking for a camp in a valley, which is at the bottom of one of the biggest valleys on earth," Mark points out, chewing on a grass stem. "It's a sump within a sump; when it rains, we're going to get wet. But we haven't found a better place. I'm willing to gamble; what about you?"

A massive line of cumulus clouds poised along the eastern scarp reminds us that we are in a race with the weather.

"We don't have much choice. What shall we name it?"

The marula trees wave their massive limbs in the gentle breeze as if inviting us to shelter here. Nearby a herd of puku is grazing along the Lubonga.

"How about Marula-Puku?"

o o o

The days of celebrated sunrises, river lunches, and goose watching are over. Each day more and more squadrons of clouds assemble to the west and north, massing like a mighty army over the scarp mountains. We must build a primitive structure at Marula-Puku and complete the all-weather airstrip on the rocky ridge upstream. The Land Cruiser will never pull through the infamous Luangwa mud, so the Frankfurt Zoological Society has sent us a new Unimog — a nine-foot-tall, six-ton, all-terrain vehicle that is a cross between a tractor and a truck. With sixteen forward and reverse gears and a low center of gravity, it can climb steep slopes without turning over. We have to collect the "Mog" from Durban, South Africa, and drive it fifteen hundred miles back to camp. All this before the rains.

Simbeye and I leave at dawn the next morning for the long trip up the scarp to Shiwa N'gandu to hire a work crew and to Mpika to buy supplies. At Mano the game guards beg me to transport their corn into Shiwa for grinding, to take one of the wives to the hospital, and to carry four scouts to Mpika to collect their pay. In the village of Mukungule one of the headmen needs a lift to Chinsali to attend a funeral. The chief's wife has to transport her bean crop to market — some say she is smuggling it to Zaire for a better price. The school headmaster needs paraffin for his stove.

By the time we leave Mukungule the old truck has eleven people on board — I'm not sure how many children — and heaves under the weight of corn, beans, two live chickens, and everyone's katundu. My daypack has only a few crackers and sardines for my lunch, not enough to share with my passengers. Not wanting to eat in front of them, I stay hungry.

At four-thirty we reach the Great North Road and stop at the few grass huts of the village of Kalalantekwe for the night. I promise to collect those going to Mpika, including the sick woman, the next day. Simbeye and I drive past the blue waters of Shiwa N'gandu Lake along a dirt track, lined with huge trees, searching for a place to camp. I ask Simbeye if he can hire five men from nearby villages to work for us, and arrange for some women and children to cut thatching grass for us to haul into the valley.

Simbeye assures me he will do this and that one of the men he will hire is an excellent cook. I have not considered hiring a cook at this stage. With so much work to do and so little food to eat, it seems rather extravagant, but I ask the name of the cook.

"His name, Madam, is Sunday Justice."

"*Sunday Justice!*"

"Yes, Madam."

"Well, bring him for sure, and four others."

After several miles Simbeye shows me a place to camp near a rushing stream sheltered by massive palms. I drive him to his father's boma a mile away, where he will spend the night. The next morning at five-thirty sharp he returns, and as we drive back toward the main road he tells me that he has hired five men, who will be ready by noon, and that his sisters are cutting the thatching grass for us. Our first stop is to deliver the scouts' corn to the local miller, who lives down a crooked, sandy road. After much negotiating over prices, Simbeye and I unload the fifty-pound bags on the doorstep of the tiny, one-room mill house. We stop on the Great North Road to collect our passengers, then drive the forty miles to Mpika, where we leave everyone at their desired destinations, and still make it to the open-air market by eight o'clock.

Women and girls, clad in multicolored chitenges (strips of cloth wrapped around the waist to make a skirt), squat behind small piles of onions, cabbages, tomatoes, rice, and dried fish spread on the ground in the center of the Mpika market. They are surrounded by cement-block stalls that offer bath soap, matches, washing powder, and a meager selection of canned foods, including the ever-present corned beef and baked beans. Spry young men — coiled to run from authorities at a shout — call out prices for black-market goods such as sugar, flour, and cooking oil. The market sways to the beat of the gumba music blaring from a small stall. It looks rather quaint and picturesque, as most third-world open-air markets do; but in this part of Africa, one man's postcard is often another man's misery. A toothless old woman tries to sell two onions, another a handful of potatoes and some dried caterpillars.

We go separate ways to buy bags of cabbages, onions, beans, ground nuts, rice, and salt. Carrying a large bucket, I approach a young man selling sugar at one kwacha a cup. Several other women, holding various containers — the lid from an aerosol can, a roll of newspaper, a plastic cup — queue up behind me. As they look from my large bucket to the shrinking pile of sugar, concern grows on their faces. I step back to let them buy first, and they bow and clap their hands in the Bemba fashion of greeting and gratitude.

At Mpika Suppliers, a general store lined with shelves of colorful cloth, basic hardware, and foodstuffs, we buy cement, whitewash, nails, and mealie-meal. There is no flour or bread at the bakery stand, and we can't find milk, honey, jam, meat, eggs, cheese, or chickens for sale anywhere in the village.

I make a courtesy call on the district governor, Mr. Siangina, at his hilltop office overlooking Mpika Village. Charming and very enthusiastic about our project, he says he welcomes any program that will stimulate the economy and discourage poaching.

Next I visit the game warden, Mosi Salama, with whom we have met on several occasions. He is built like a bowling pin with a cheshire cat smile and eyelashes as long as a moth's antennae. Mosi greets me on the veranda of the lime-green concrete-block building that is the Mpika headquarters for the National Parks and Wildlife Services. He assures me again, smiling broadly all along, that he did issue the ammunition and food to the Mano scouts. When I inquire further, he gives me the surprising news that an officer from the Division of Civil Aviation, Mr. Banda, and the National Parks pilot, Captain Sabi, have driven all the way from Lusaka and are sitting in his office waiting to see me. As soon as I enter, without shaking my hand Mr. Banda says, "I'm afraid there is a very big problem with your program. You have been landing your airplane in North Luangwa National Park without permission from our department."

Immediately I relax. "Oh, it's okay. See, I have the papers right here." I dig in my briefcase, pull out a thick file of permits, and shuffle through them. "Here is the permit from the Zambian Air Force giving us permission to operate our plane in North Luangwa. And this is the permit from the minister of tourism granting us permission for our project, which as you can see in this paragraph, explains that we will be flying an aircraft in the park. This is a photocopy of Mark's Zambian license and a three-month blanket clearance for operating in this area from your own department."

Captain Sabi and Mr. Banda lean their heads together to read the documents. After a moment Mr. Banda shakes his head. "There is no permission here to land your airplane in North Luangwa National Park."

"Operating surely includes taking off and landing." I smile and try to make a joke. "Did they think we were going to fly forever and not land?"

"You cannot land an airplane in Zambia on an unregistered airstrip. You will have to operate from Mpika," Captain Sabi informs me.

"Mpika airstrip is fifty miles from our study area. How can we operate from there? We plan to build a proper airstrip. In the meantime, as we informed the ministry, we have cleared some temporary strips." I try to make another joke. "Mark's been a bush pilot for years; I'm not sure he could land on a real strip." No one smiles.

"You cannot land your plane anymore until you have constructed an airstrip that the DCA approves," Mr. Banda insists. "It must have concrete markers, a wind sock, all the requirements."

I slump in my seat. We have spent more than a year obtaining permits. This new requirement will set us back months, for we will be unable to conduct aerial surveys or antipoaching patrols until we have completed the strip. I look at Mosi to appeal for help. He appears thoroughly disgusted, but whether at me or at these men I can't be sure. He gives me no support whatsoever.

"All right. Thank you." I stand and walk out of the room, managing a slight smile when I imagine what Captain Sabi and Mr. Banda would think of Airstrip One, with its short runway and looming sausage tree.

o o o

There is nothing quite like the sensation of returning to a bush camp. I don't know whether it is more exciting to be the one waiting in camp, listening for the distant drone of the engine, or to be the one coming home. In the silence of the wilderness, the one in camp can hear the truck approaching from so far away that there is a sort of meeting of the hearts before the other arrives. As I pass the mopane tree, the one with the marabou stork's nest, I know that Mark can hear the truck, and he knows that I know. The rest of the journey is like an extended hug, so by the time I roll into camp, we are both warm and smiling.

Around the campfire we talk about my trip, until I cannot avoid the bad news any longer. When Mark hears that we cannot operate the plane in the park until we have a proper strip, he is upset; but we vow to work even harder to get the strip completed before the rains. Meanwhile, we just won't fly.

The next morning Mark drives the Cruiser and trailer, loaded with building materials, from our little camp to the Marula-Puku site. He, Simbeye, and the other men begin building the mud-wattle hut that will be our shelter in the rains. Our new workers are young tribesmen in their early twenties, dressed in ragged Western clothes but without shoes. They have never held a job except for helping their fathers tend the tiny maize and groundnut patches on their subsistence farms. The only tools they know how to use are axes, hoes, and shovels. But while they have had no more than five years of schooling, they are eager to learn. Of the five new men Sunday Justice, Chanda Mwamba, and Mutale Kasokola in particular radiate good humor and an uncommon willingness to work hard, as does Simbeye. Today I had said, "Good morning, guys," and they had laughingly repeated the word "guys" over and over. The name stuck and they call each other "guys" to this day.

Sunday Justice, bright-eyed, short, and chubby, stays with me to unpack the supplies. We store beans in a large can, tissue and soap in a large blue trunk, and sort through the other groceries. As we work, I explain to him that we hope to save the park by making it benefit the local people; we may even hire some of the poachers to work with us.

"That is a very good idea, Madam. You should go to Mwamfushi Village, where there are many poachers."

"Tell me, Sunday, can we fly to that village?"

"Oh no, Madam, that village is very much on the ground." I smile behind his back for a long moment.

All morning I have noticed Sunday stealing glances at the plane, parked in its thorn boma.

"You like the airplane, don't you, Sunday?"

"Yes, Madam. I myself always wanted to talk to someone who has flown up in the sky with a plane."

"Well, you can talk to me," I say, as I pour salt into a jar.

"I myself always wanted to know, Madam, if you fly at night, do you go close to the stars?"

I explain that on earth we are so far from the stars that being up a few thousand feet does not make any difference in how close they look. But I don't know if he understands, so I end by saying, "When you fly at night, you *feel* closer to the stars."

When we have finished unpacking and organizing camp, I show him our few kitchen utensils and try to discover how much he actually knows about cooking. He can bake bread, but only in an oven, he tells me. It seems that I will have to teach this African how to cook in a traditional black pot. In our wooden bowl I mix the ingredients for cornbread — omitting eggs and butter, of course, because there aren't any — and show him how I bake it in the pot, putting hot coals on top and underneath.

"Now this is something you have to be very careful with," I say as I hold up a frying pan with a nonstick coating. "See, it is not an ordinary pan. Food won't stick to the bottom. It's like magic."

"Oh, a Teflon pan," he nods. I stare at him. How often do you find a man who knows about Teflon, but wonders if you can fly to the stars? More often than you would imagine, I decide.

All morning the sun shoulders us with heavy heat, but to the west a blanket of dark clouds is draped across the mountains of the scarp. Our temporary base camp is on the other side of the river from Marula-Puku, about fifty minutes downstream. If it rains in the mountains, the river could flood, making it impossible to cross until the dry season. After giving Sunday a few more chores to do, I wander down to the riverbank to see if the water has risen in the last few days. It is hard to tell, so I kneel down in

the sand and lay a line of small stones at water level. Each day I
will check the stones to see if the water is rising. We cannot yet
camp at Marula-Puku; we must stay near the plane to guard it
against hyenas and poachers.

"What are you doing, Madam?" Sunday has come up quietly
behind me.

"Well, Sunday, I need to know if the river is rising, because we
will have to leave this camp and move to the other side before it
gets too high. I can measure the water level by these stones."

"It won't be like that, Madam."

I stand up and look at Sunday. "What do you mean?"

"Today this river will be here." Sunday points to the water
lapping against the sand by our feet. "And tomorrow when the
rains come, this river will be there." He points a hundred yards
across the floodplain to a spot far beyond the camp and airplane. I
look back and forth from the river to the plain. If the river rises
that high, the Cessna will float away like a raft.

"Sunday, you mean the water will come up that far in one day?"

"Maybe one hour. If these rains come to these mountains, the
water, she comes down all together." As if on cue, the clouds to
the west release a slow, distinct thunder, which rolls heavily over
the hills.

"Sunday, come with me." Back in camp, I begin flinging boxes
open and sorting through gear. "Take everything I give you and
put it in a pile under the tree." I separate daily essentials — food,
clothing, cooking gear — from valuables — camera equipment,
film, tools, spares. Pointing to the latter, I say, "We'll have to carry
this stuff to Marula-Puku."

At daybreak the next morning, Mark and I transport all of the
valuables to Marula-Puku camp, which is on higher ground. Leaving the other men at work on the Bemba house, we head down the
Mwaleshi to collect the cache we hid last year. I dig out our
handmade map that shows how to find it: "Drive 15.6 miles along
the north bank of the river to our Palm Island campsite; go 1.3
miles north through the trees, past two large termite mounds;
cache is hidden under *Combretum obovatum* bush."

In spite of the dense October heat and our race with the rains, the journey along the Mwaleshi is as wondrous as ever. There is still no track, but we know the route now and avoid the worst sand rivers, the deepest gullies, the muddiest lagoons. Puku, waterbuck, zebras, and impalas graze on the broad plains as we drive along. A herd of more than a thousand buffalo shake the earth as they canter across the grassland. White egrets, stirred by the stampede, soar like angels against the black beasts. Small bushbuck peer at us from the undergrowth with Bambi eyes.

After only thirty minutes of slogging across dry riverbeds and small savannas, we approach a very familiar plain. We turn inland toward a stand of *Trichilia emetica* trees hanging over a bank. I touch Mark's arm and he stops the truck. Up ahead, under the same trees where the cubs played last year, lies the Serendipity Pride. Two adult males with flowing manes, three adult females, and four stroppy yearlings stare at us from the grass. The young female, the one we called Serendipity, stands and walks directly toward us, her eyes not leaving us for an instant. The males, apparently still shy, slink away in the tall grass. The four cubs trot behind Serendipity toward the Cruiser, and once again in this old truck we find ourselves surrounded by lions.

Serendipity stares at Mark from two feet away, then smells the front wheel for a long moment as the Blue Pride lions had done so many times. The four cubs — all legs and tails — soon tire of this odd beast with its strange odors of oil and fuel, and bound away in a game of chase.

We long for the day when our camp will be finished, so that we can spend more time with the lions. It will be fascinating to compare the social behavior of these lions with those we know in the desert. Forget the cache. Flash floods or not, tonight we stay here to watch the lions.

In the late afternoon they rouse themselves, stretch, and stroll toward the river. Serendipity leaves the others and walks across a small plain to a ten-foot termite mound that towers above the grass. With one graceful leap she springs to the top of the mound and balances on all fours. Her tawny coat blends with the gray clay, as she searches through the grass for any signs of supper.

After only a few moments she slithers down the mound in one silent motion and stalks through the grass toward the riverbank. The other lionesses raise their heads to full height, watching Serendipity's every move. The cubs stop their play and look at the females. The bank drops down about eight feet and Serendipity jumps out of sight onto the sandy beach. At that moment a male puku emerges from the grass, his black eyes searching the beach. Serendipity vaults over the bank to within five feet of him. In one motion she springs forward, reaching out a powerful paw and tripping her prey. At the same instant the other two lionesses dash from their positions just in time to pounce upon the struggling ungulate.

In seconds the three adults and four youngsters have their muzzles into the open flesh. The lionesses feed for only a minute, then walk away licking one another's faces. A hundred-forty-pound puku doesn't go very far among seven lions, and they leave the rest for the cubs.

Serendipity and the two other lionesses walk across the white beach and lie next to the water's edge. On the far bank a small herd of buffalo hold their muzzles high, snort, then disappear into the golden grass of sunset. Beyond them a herd of zebra grazes unaware. The lions will have to hunt again tonight, but prey is plentiful and water laps at their toes. We can't help thinking of the desert lions, who lived for two years eating prey as small as rabbits and had no water at all to drink. What would happen if the Kalahari and Luangwa lions traded places? Could these lions survive in the endless dry dunes of the desert?

After dark all seven lions cross the river in a long line, and we camp near their *Trichilia* tree. In the morning, when there is no sign of the pride, we continue our trip down the Mwaleshi to recover our cache.

Tall grass has grown up along the base of the bush, so the cache is even better hidden than when we left it. We cut the thorn branches free and pull out drums of diesel, jars of honey, jam, and peanut butter, even canned margarine from Zimbabwe. We sling the winch cable over a tree limb and load the drums into the trailer. After a not-so-refreshing swim in the hot river, we begin the long haul back to Marula-Puku.

o o o

Making endless trips up and down the jagged scarp — to collect poles for the hut, to haul more thatching grass, to meet Mark after he flies the plane to Mpika — we race to finish Marula-Puku camp. The heat and flies must be feeding on each other; both have become fat and unbearable. When the hut's mud walls reach eye level, it becomes obvious that, although their spirits and intentions could not be better, the Bembas are not the builders they claim to be. One corner of the hut crumbles before it has dried and we are doubtful the house will withstand the rains. But it is too late to start over, so we smear the cracks and holes with gooey mud and tie bundles of grass to the roof.

Afraid to stay in our little camp on the far side of the river any longer, we move to Marula-Puku one morning in early November. I am greatly relieved to have all of the gear at one campsite. While Mark joins the hut-building crew, Sunday and I set up our new temporary camp nearby. Sick of making camps, I hope this will be the last one for a while. Once more Sunday and I store the blue trunks in the tent, hang the cornmeal in the tree, put up the tables and chairs. As we work, white-browed sparrow weavers — the same species that shared our Kalahari camp for seven years — serenade us from a wispy winter-thorn tree (*Acacia albeda*). In the lower branches of the acacia they too are busy building their nests, but unlike other species of weavers, these seem unconcerned with neatness. They twist grass stems of various sizes and shapes into a messy bundle that looks like something that might be cleaned out of a drain. But their familiar song and perky chirps lighten my heart. If they can sing while they build, then so can I.

By midafternoon we have all the gear set up at Marula-Puku and I walk to the other side of the island to help the men with the hut. As I kneel in the sand, spreading mud with my fingers, I feel the wind pick up and stir the muggy air around us. A cloud mass looms to the northwest, but such formations have been marching harmlessly overhead for days. Suddenly a hot wind rushes into the valley, ahead of a low, swirling wall of black-gray clouds. Jumping to our feet, we grab gear and supplies and throw them into the unfinished hut. Mark and I run toward the tent, but lose sight of it in the dust and sand. Finally, we stumble over it lying on the ground. Pots and pans, books, chairs, plates, cups, and clothes are scattered for yards around. The sand stings our faces and eyes as we stagger through the storm, chasing bits and pieces of camp and throwing them into the truck.

The sandstorm lasts for thirty-five minutes and then retreats as quickly as it came. Not one drop of rain has fallen, and the valley is left standing in a still, dry heat. Silently we wander through the remains of our camp, standing up a table here and a chair there. The sparrow weavers are chirping furiously — their nests have been blown to the ground — but the next morning when we rise, they are rebuilding with crazed enthusiasm. I watch them briefly and do the same.

The ominous date of November 14 — when according to British meteorologists the rains begin — is less than two weeks away, when Mark and I drive to the top of a wooded hill to lay out the airstrip. We have chosen this site because it will drain during the rains, making it usable all year. This will be an easy job, I think, but the woods are so thick that it is impossible to see ahead far enough to lay out a straight line for the runway. After pacing off fifty yards, we run into a huge termite mound, so we try another heading and run into a deep gully. Deciding we must do a proper survey, we get the compass and hammer a row of three stakes into the ground every fifty yards, one in the middle of the strip and one on each side.

About this time a greater honey guide, a small gray bird with a black throat, notices us and gives his distinctive "chitik-chitik-chitik" call, inviting us to follow him to honey. Honey guides lead humans to bees' nests, and when the hive is opened for honey, the birds eat the bees, the larvae, and the beeswax. It is not at all difficult to follow a honey guide. He gets your attention by flitting about in a tree near you, making his raucous call. As soon as you approach, he flies off in the direction of the bees' nest. And just to make sure that you are still there, he stops often in trees along the way, calling over and over.

The honey guide at the airstrip sees us walking directly toward him and flies about in excited, exaggerated twirls. In this unpopulated area, he can't very often have a chance to lead people to honey, and he is more than ready. Just as he is about to fly to the next tree, we hold up our compass, turn a sharp ninety degrees, and march off in another direction. He is silent for a moment, then flies to a tree in our path and calls even more vigorously than before. Again we are nearly under his perch when we swing ninety degrees and pace off fifty yards in another direction. The airstrip is to be more than a thousand yards long, so forty times we stalk across the woods in one direction, only to change course and prance off in another. The poor honey guide has obviously never seen such stupid people. He takes to flying right over our heads, flapping his wings wildly. Now and then he lands on a limb nearby and glares down at us with beady eyes.

Sweat mats our clothes and grass blades sting our legs as we trundle along through the trees. Thinking it was going to be an easy job, we did not bring enough drinking water and our throats are burning. Finally, we hammer the last stake into the ground and climb into the truck. As we drive away, the honey guide is still perched on a branch, giving an occasional "chitik-chitik" with what seems a hoarse and raspy voice.

o o o

The Bemba hut, with its scraggly grass roof and big windows framed with palm stems, looks more like a lopsided face with a crooked straw hat than a house. But because we must, we declare it finished. At five-thirty the next morning we are all at the airstrip site, armed with six axes, six hoes, three shovels, and the truck and chains. We have about three thousand small trees to cut down, and three thousand stumps to dig up. The ground will have to be leveled and graded by hand. It is impossible to know where to begin, so we just start slashing, chopping, digging, and shoveling. Two hours later, the temperature is already above 90°F. Mark and I wet our clothes from the jerry can with river water that smells strongly of fish and buffalo dung. The Bembas, born of this heat, smile at this. They are not yet hot. "Wait until noon," they laugh.

After a while a routine emerges. Mwamba, Kasokola, Sunday, and I cut down the small trees and pull them to the side; Mark and Simbeye follow and yank out the stumps with the chain and truck. The others chop out stumps too large for the truck and fill in holes. In the midday sun the seventy-five-pound chain is so hot I cannot hold it, but all day long Simbeye runs barefooted from one stump to the next, carrying the burning links across his bare shoulders, refusing to let anyone relieve him or even to rest. As we work, the guys sing softly in a language I do not understand, but with a spirit I certainly appreciate.

To the south we see a flat-topped mushroom cloud rising from the parched earth — the poachers are setting wildfires that burn unchecked across the park. No doubt they are also shooting elephants for ivory and buffalo for meat, yet the game guards have not mounted a single patrol. Without roads or our plane there is nothing we can do to stop the killing and burning. I hack with rage at a small bush; we *must* finish this airstrip.

Where is the rain? For days we have worried that it will come; now we worry that it won't. Stranded or not, flooded or not, anything is better than this heat.

Each day we work from 5:00 to 11:00 A.M. and from 2:30 to 6:00 P.M. Some afternoons I stay in camp by myself, hauling water from the river in buckets, boiling drinking water, baking bread, collecting firewood, and washing clothes. All that is left of the Lubonga River is a few stagnant, smelly pools that we gladly share with the buffalo, puku, and zebras. They come late every afternoon to drink, and frequently I collect water with buffalo standing nearby. They seem to have lost their inhibitions — and so have I, for they used to be the one animal that really frightened me. Shrinking resources make either fierce enemies or strange friends.

At night Mark and I sleep on top of the truck, where it is cooler than inside the hut. Still, the temperature is often more than 100° at midnight, so we lie under wet towels trying to stay cool. We awake just before dawn to watch the stars retreat into the brilliant colors of sunrise and to see the water birds — saddle-billed storks, yellow-billed egrets, Goliath herons — soar along the river on their way to work. Commuters, we call them. They do not take shortcuts across the bends in the river, but follow its winding course, maybe for the same reason that we would — to get a better view.

One morning, while hacking at a small tree on the western end of the airstrip, I am surprised to see a straggling line of game guards walking toward us, some in tattered uniforms, others in civilian clothes. Nelson Mumba still has a red bandana around his head, Island Zulu carries the bent frame of a cot, Gaston Phiri totes a live chicken under his arm. Mark and I greet them and explain that we are building an airstrip.

"We'll use the plane to spot poachers and to count animals from the air," Mark says.

They nod and he continues. "In fact, we were thinking, if you would help us build an airstrip near your camp, we could use the airplane to take you on patrols in the park."

"Ah, but we cannot build an airstrip," answers Mumba.

"Why not?" I ask. "We'll lend you some tools." I hold up my ax.

"We are officers; we do not do manual labor," Mumba says, looking to the others, who nod in agreement.

"Not even if it makes your lives easier?"

"We do not do manual labor," he repeats. Mark, sensing my rising anger, changes the subject. The new subject is not much better.

"Are you going on patrol?" he asks.

Two of the scouts answer "yes," three say "no."

"Are you going to shoot for meat?" I ask.

"We have the right, if we want to," Mumba answers in a surly voice.

Mark explains to them, although we are sure they already know, that according to government regulations no one, including game guards, can shoot animals inside the park.

They talk to one another in Chibemba, their eyes darting back and forth at us. Mark adds that if they hunt in the park, we will be forced to report them to the warden. This is not the way I want this conversation to go; we want to work with the guards, to encourage them. I try to think of some way to salvage the situation.

"Look," I say, "we know it's rough living in your remote camp. When we get our project going, we want to work with you, help you. We'll buy you new uniforms, camping equipment, things like that. But in return you must do your jobs. You're hired by the government to protect these animals, not kill them!"

They talk some more in Chibemba and then announce that they must be going. We shout friendly farewells — "Good luck on your patrol" — but these words bring no response and wither in the heat.

Mark and I stand, watching the scouts hike over the hill. We disagree about how to fight poaching. He believes that we should get personally involved — flying patrols, airlifting scouts, going on antipoaching foot patrols with the guards. I argue that we should supply them with good equipment and encouragement, but we should not personally go after the poachers, for then they will come after us. Unarmed, we make an easy target. We have discussed these points over and over, but have not come to an agreement.

"Better get back to our manual labor," Mark jokes.

"Sorry, I'm an officer," I say as we both begin shoveling again. We've been working on the strip for over a week, and while half of it is clear of small trees, it still looks more like a spot where elephants have romped in the woods than an airstrip.

Two flat tires on the Cruiser force us to return to camp early in the afternoon. When we arrive, Tapa, the game guard enlisted to protect our camp, is nowhere to be found. Mark and I find him hidden in the tall grass, drying meat on a rack. Every day, while we have been clearing the airstrip, Tapa has been fishing, trapping, and drying meat. Stunned, we ask what he is doing, and he replies calmly that he is going to sell the meat in the village. "That is against the law, you know," I say. He shrugs his shoulders and continues to poke at the charred body of an otter, sizzling over the glowing coals. Although we do not report Tapa, we send him back to his camp without the food, and never again do we employ a game scout to watch our camp. It is beginning to seem that instead of the guards protecting the park, we need to protect the park from the guards.

While the men are busy mending the tires, I haul water and bake bread. Several times during the last few days it has rained high in the scarp mountains, and our little river is flowing gently with a few inches of water. We do not know when the real floods will come, so I have been pestering Mark for days to move our valuables again — this time to the airstrip, which is on still higher ground. Most of our gear is stored in the hut, but I'm worried that with the first rain it will be reduced to the mud from whence it came.

Bending over a basin washing clothes, I hear a distant roar and stand up to listen.

Simbeye and Mwamba shout, "The river, watch the river. It comes!"

Instead of running away from the riverbank, which would seem the sensible thing to do, we all run toward it and look upstream. A wall of water, three or four feet high, rounds the bend to the north. It rushes toward us, mad and muddy, looking like one river flowing on top of another. We watch in disbelief as the Lubonga, which was only five yards across, widens to more than a hundred yards, spreading over the rock bar on the opposite side.

Our bank had been a good twelve feet above the river; now the rushing water is only four feet below us, and rising. Huge trees and branches bob in the waves.

I scream over the roar, "Mark, it's going to come over the bank! We've got to move everything!"

"Let's wait and see," he shouts back. I wonder if we will stand here until our feet are wet before we do something.

"Here comes another river," Simbeye calls, pointing behind us, and we run to the northwest corner of the island. Savage brown water pours from what had been a dry gorge. It fills the oxbow to the north of camp with an instant roaring river thirty yards wide. Swollen rapids, topped with foam and debris, cover our track. We are cut off, stranded on the island. There is no longer any chance of moving to the strip. We run back to the main river, which has already risen another foot. Why haven't we listened to all the warnings about floods?

"Shouldn't we at least put everything on top of the truck?" I plead with Mark.

"I don't think it's going to come any higher."

"How on earth do you know?"

"Don't worry. It's going to be okay."

The seven of us stand on the riverbank, heads down, watching the unleashed fury of our little river. A large chunk of the bank falls into the hungry current. Mark motions for us to move back from the edge. The water is two feet below the top of the bank, but its rise seems to have slowed. Ten minutes later it is unchanged.

"See, it's going to be okay," Mark says.

"Right, fine, we're not going to be washed downstream," I agree. "But now we're stuck here; we won't be able to drive out for weeks."

"Oh no, Madam, it won't be like that," Sunday smiles. "This water, she will be gone tomorrow."

"The flood will only last one day?"

"This water," he explains, "she is coming from the mountains to the Luangwa River. When she is there, she will not come back. More water will come on a day from here, but this water, she will be gone tomorrow."

"See," Mark smiles, "no problem. This water, she will be gone tomorrow."

o o o

The flood begins to recede after a few hours, but it is only the first of many floods to come. With each rain, whether here or in the mountains, the rivers will become higher, the ground soggier, the track more slippery. The old Land Cruiser will no longer be able to drive us out of the valley. We must finish the airstrip and collect the Unimog truck, or we will not be able to operate in North Luangwa this season.

For now we are stuck in camp; we can't even drive to the airstrip until the flood recedes farther. Mark and I pull our folding safari chairs close to the riverbank and watch the Lubonga with new respect. She is still raging, and now that we know her moods, we will never take her for granted.

At sunset a small group of puku females gather on the opposite bank and stare at the river; one of their favorite sleeping spots is three feet underwater. After a while they settle down in a tight knot in the grass.

Darkness brings dazzling stars and the first lightning bug of the season. In a few days thousands of fireflies will sprinkle the night with their phosphorescence, like sequins fluttering and floating through the balmy air. But tonight this one is all alone, and seems lost and lonely as he flashes unanswered valentines above the grass. Usually lightning bugs fly no higher than the treetops. But this one soars higher and higher toward the starry sky as if, for lack of a mate, he has fallen in love with a star.

o o o

In a final dash to finish the runway before the rainy season, we start working on it at four-thirty every morning. As with most of our races with the African elements, we are losing. Nearly all the small trees have been cleared from the airstrip, but hundreds of stumps remain to be pulled, and three termite mounds as hard as concrete and as large as the truck must be leveled.

Now that much of the undergrowth has been removed, some of the animals can't resist the lush green herbs and grass that have been exposed. A male puku with one horn has already claimed part of the airstrip as his territory, and has become so habituated to our presence that he grazes nearby as we hack and chop. A family of warthogs and a small herd of zebras often forage at the opposite end of the strip.

One hot afternoon, as we are lost in a haze of heat and work, Simbeye calls softly, "Nsofu, there." Far on the other side of Khaya Stream, in a little valley, we see ten elephants moving through the tall grass. They are the first living ones we have seen since our return to North Luangwa. We watch them in awe and whisper softly when we speak, even though they could not possibly hear us at this distance. Adult males, three of them without tusks, feed on the prickly branches of the winter-thorn trees. Reaching with his trunk, one of them pulls down a branch of a fifteen-foot acacia. He strips the bark — twigs, thorns, leaves, and all — and stuffs it into his mouth.

We see these elephants on several more occasions, always in the distance, and although it seems a bit of an exaggeration, we start calling them the Camp Group.

November 14 arrives, ripe with legendary promises, but still there is no rain. Sunday was right, the river has become a gentle stream again, and only the driftwood high on its banks tells of the flood. But every afternoon giant cloud formations rumble across the sky. Simbeye keeps telling us that we must go now, that once the rains come we will not be able to drive up the slippery mountain track in the old Cruiser. "Only one more day," Mark keeps saying, as we pull more stumps and fill more holes.

Finally, the airstrip resembles a runway, although it is certainly not yet ready for approval by the Division of Civil Aviation. There are still several large hummocks, where termite mounds used to be, and scores of stump holes. But it is close enough to being finished that we can complete the job quickly in the Unimog when we return from Durban.

With no idea how long we will be away, we pack up all the gear and hide another cache of leftover fuel and avgas near the airstrip. With all the Bembas bouncing in the trailer, we drive up the scarp. We leave them at Shiwa N'gandu, with the promise that we will hire them again as soon as we return — in a few weeks, we hope.

We are near Mpika when fat raindrops pound the top of the truck, and curtains of white rain drift through the air. We look back. The valley, filled with cumulonimbus clouds, looks like an enormous bowl of popcorn. We have made it out just in time.

7
A Valley of Life

DELIA

> The most present of all the watchers where we camped
> were the animals that stood beyond the firelight,
> being dark, but there, and making no sound.
> They were the most remembered eyes that night.
> — WILLIAM STAFFORD, "When We Looked Back"

o o o

CARRYING THE HOT-WATER KETTLE, a towel, and a flash-light, Mark follows the footpath through the dark camp toward the bath boma. Surrounded by tall grass, the boma is a three-sided structure of sticks and reeds standing at the edge of Marula-Puku camp. Inside is a wash table also made of sticks, a jerry can of cold water, and a basin in which we mix hot and cold water for our baths every evening.

A rustling noise sounds from the grass. Mark pauses briefly, but walks on. Earlier a male waterbuck had been grazing near the sausage tree just beyond the boma; he is probably still in the tall grass. Mark mixes his bathwater, then switches off the flashlight to save the batteries. Standing naked under the bright stars, he begins to wash his hair, closing his eyes against the soapsuds. He freezes as he hears the rustling again from six feet behind him, just outside the boma. He quickly splashes water onto his face to rinse away the soap, then switches on his flashlight, hoping to get a closer look at the waterbuck. A wall of tall grass is all that he sees, but he can still hear the swishing sounds. He steps to the grass, parts it, and shines the light into the thick cover at his feet.

A lioness, crouched flat against the ground, glares back at him, her tail lashing. She is only four feet away, looking straight into his eyes.

"Aaarrgghh!" Involuntarily, Mark utters a primal growl and

jumps back. At the same instant the lioness springs to her feet, hissing and spitting at him, her canines gleaming white in the light. Mark leaps into the boma. The lioness whirls around and charges away through the grass to join five other lions ten yards behind her. Together the pride trots to the firebreak a little farther on, where they all sit on their haunches, staring at Mark. Suds dripping down his neck, Mark stares back, wondering if they will become frequent visitors to our camp like the lions we knew in the Kalahari.

o o o

It is early February 1988, much later than we'd planned, before we return to Luangwa — Mark in the new Unimog hauling a thirteen-ton shipping container, I driving the old Cruiser. The rains have transformed the Northern Province into a lush tangle of weeds, grass, and shrubs. The little villages are smothered in vines, dripping with today's rain and yesterday's moist blossoms. To stay dry the women have to cook inside their grass huts, the smoke smoldering through the thatched roofs. When we reach the village of Shiwa N'gandu, Simbeye, Kasokola, and Mwamba rush out of their mud and thatch huts to greet us. All smiles as usual, they are ready to return to Marulu-Puku. When I ask about Sunday Justice, they explain that he has gone to Lusaka to look for work. The thought of this gentle, soft-spoken, imaginative fellow walking the tough streets of Lusaka saddens me. But, happy to have the other men with us, we start once again down the scarp.

Island Zulu, Gaston Phiri, Tapa, and the other game guards leave their beer circle and shake our hands warmly in welcome. The children surround us calling "smi-lee, smi-lee," which at first I mistake as some form of Bemba greeting. Then I realize that I have always asked them to smile for the camera, so they think that "smile" means "hello." When I greet them in Chibemba, they collapse in giggles and run into their huts.

The Mwaleshi, now in full flood, does not give us such a warm welcome. The swollen river tears through the forest, splashing spray and whitecaps against boulders and logs. If we cannot get across the river, our hopes of working in North Luangwa during the rains will stop right here.

Using the bucket loader on the Mog, for three days we quarry stones from a rocky outcropping and dump them into the river to make a ford. On the morning of the fourth day Mark ties a rope around his waist and swims across, pulling the end of the winch cable with him. Once on the other side, he hooks the cable around the base of a large tree and recrosses the river. As he slowly eases the nine-foot-tall Mog into the river, its hood disappears under the roiling current. The cab rocks wildly as the truck climbs over the boulders on the river bottom. Water seeps in around Mark's feet, but the truck churns through the current and pulls itself up the muddy bank on the east side of the Mwaleshi.

After several more trips to ferry the rest of the gear across, we fill the back of the Cruiser with large stones for ballast and winch it over. Water pours out of every door and crack of the old machine, which must long for desert days gone by. The current would sweep away the shipping container, so we leave it in Mano.

On our way down the scarp the trucks get stuck so often in the greasy mud that we don't make it to Marula-Puku tonight, as we had hoped. We camp near "Elephant's Playground," a swale of long grass where we have often seen elephant tracks, dung, and broken trees, but never the elephants themselves. The next morning the grass is so tall — ten feet in places — that it is difficult to follow the track. Simbeye, Mwamba, and Kasokola climb into the Mog's bucket, and Mark raises them high over the truck so they can guide us.

The runway we had worked so hard to clear is covered by grass eight feet tall and hundreds of small mopane shrubs. "It's not as bad as it looks," Mark tries to cheer me. As we continue down the track, I dread what we will find at camp. From the tall north bank, all that can be seen of the Bemba hut is the soggy, lopsided thatch roof that appears to be floating on top of a grass lake. The oxbow around the island is a swamp with tall reeds bobbing in knee-deep water. But the main river, the Lubonga, is well within its banks, and only a shallow stream separates us from camp.

The adobe walls are cracked and crumbling, the roof wind-blown and certainly not waterproof, but the hut is still standing. As I walk silently inside, Simbeye says, "This is not a very fine house, Madam, but we will make it strong for you." I smile grate-fully at him and look around. One big truck, one small truck, and five muddy people — a motley crew for the size of the project we have in mind. I look up at Mark. "Well, we'd better get started."

There is plenty of willing mud to repair the walls and plenty of ready grass to mend the roof. Within two days we have a primitive camp and a week later the grass has been cleared from the airstrip. The runway still has to be leveled, which will take another three months of backbreaking work, but after that we can fly the airplane to the valley.

Around the campfire each night, Mark and I talk endlessly about how North Luangwa can be saved from poachers. We will have to start by working with the game guards to enforce the laws against poaching, but that will be only the first step. The people who live around the park must be convinced that wildlife is more valuable to them alive than dead. Eventually, we hope that conser-vation-minded tour operators will run quality, old-fashioned walk-ing safaris that will put money in the pockets of the local people. The government has agreed that 50 percent of the revenue from tourism in North Luangwa — once it starts — can be returned to the villagers.

Of course, we would rather North Luangwa be free and wild, but that is no longer a choice. It lost its freedom when the poachers fired the first shot. The challenge is to save it without breaking it of its wildness.

It will be some time before revenue from tourism will begin to flow to the villages near the park. In the meantime, we can help the people find other ways of making a living so they can give up poaching. We think of cottage industries, such as carpentry shops, beekeeping, maize mills, and sunflower presses. We can help them grow more of their own food, especially sources of protein such as poultry, fish, beans, and peanuts. Hungry people do not make good conservationists.

Most important, perhaps, we will start teaching the young

people that wildlife is the most valuable resource in their district. Most of the children have never seen live elephants, much less thought of them as anything other than a source of meat or ivory. Neither has the rest of the world. This is as good a place as any to start.

Then, of course, we have to learn more about North Luangwa, especially its wildlife and ecology. We will fly regular aerial surveys, taking a census of each wildlife species and noting its distribution in order to determine whether the population is stable or declining. We will also explore the park from the ground, but that will be difficult at the height of the rainy season. Even the Mog bogs down in mud between camp and the airstrip.

So, hiking in misty mountains and across soggy savannas, and wafting in the plane over the backs of buffalo, elephants, zebras, and wildebeest, we begin discovering this land of rivers and this valley of life.

o o o

"We're not going to make it tonight. Let's camp here." Standing in the drizzle, Mark, Kasokola, Mwamba, and I look up at the sheer forested cliff towering between us and our destination. Days before in the airplane, we had followed the Lubonga River from its source on the plateau above the Muchinga Escarpment. Beginning as a narrow trickle, it wound its way through the lush vegetation and rounded peaks of the tumbledown mountains. Now and then it cascaded over boulders, creating waterfalls hidden beneath tropical trees, ferns, and vines. At one spot, near the base of the mountains, the river surged through a small, pasture-like floodplain, the shape of a teardrop, tucked away in the folds of the steep-sided hills. We called it Hidden Valley. At this point along the river's course, a single mountain ridge blocked it from the plains beyond. But decades before, the Lubonga had found a weakness in the strata of the ridge and had crashed through, creating a narrow chasm, covered in ferns, vines, and sprays of bamboo.

Determined to reach this vale, we have hiked from camp along the Lubonga for two days through mist and torrential downpours.

By keeping records of the species of wildlife and plants we encounter in each habitat, we begin to understand the flux and flow among the plant and animal communities in the valley. Thinking that we would easily make it today, we have walked until late afternoon along the Lubonga. But from the base of the last rock outcropping between us and Hidden Valley, we can see that the route is more formidable than we thought from the air. Night will fall before we reach our destination.

"Okay, this is a great spot to camp," I agree, looking around at the lush plain that stretches to the small mountain. The Lubonga, in full flood, rushes through the narrow canyon and then meanders across the soggy bottomland. I lift the heavy backpack from my shoulders and ease it to the ground.

"Look — buffalo?" Kásokola points northwest toward a low stand of *Combretum* trees. Stooping over, Mark and I look under the scattered trees to see three hundred buffalo milling about in the mist, not a hundred yards from us. Some are bedding down in the soft grass for the night; others continue to graze toward us. Apparently they have not seen or smelled us, and as we set up our little fly-camp, some of the herd wander still closer, grunting and mooing until at last they lie down, very near our camp.

As Mark pulls the tent's guy rope, he looks over his shoulder, then points to the buffalo. We all stop in our tracks. A few large females have stood up and are blowing their alarm call as they stare at us. Suddenly the entire herd are on their feet. Blocked by the ridge, the river, and the steep canyon walls, they have nowhere to go except in our direction. Their feet sloshing in the soggy ground, they stampede past our campsite, then disappear in the gray swirling mist.

Up before the sun, we break camp and walk to the gorge. Our plan is to follow the river through the canyon to Hidden Valley on the other side of the four-hundred-foot ridge. But the river is so high that it crashes against the steep walls, leaving no space for hiking. We will have to climb over the ridge.

Following an elephant trail up the hillside, we pause here and there to clip samples from towering trees and tropical undergrowth for our plant presses. Water from the leaves drips down my neck, and my last pair of dry socks is soaked. When we reach the top, Hidden Valley, nestled in its own secret hills, lies quietly below us. Shrouded in a veil of mist, giant trees and bamboo line the tiny, grassy valley — only a few hundred yards wide. The Lubonga snakes gently through the marshy bottomland, and small herds of puku, buffalo, and waterbuck graze the lush, green foliage on the banks.

For the next two days we trek through the lofty forests and across the small dambos — sunken, grassy glades — recording animal and plant life. We see tracks of the rare sable antelope and come upon a family of wild pigs digging for roots. Elephant paths worn inches deep in the hard gravel of the ridges suggest that they have been traveled for centuries by generations of pachyderms winding around the hillsides.

One afternoon, leaving the Bembas in our little fly-camp, Mark and I follow the meandering Lubonga through Hidden Valley. We cross the meadow and follow the river up to where it springs from the hills. Here the heads of the tall elephant grass rustle as we pass, and in the distance the scarp mountains brood. We step into a small, marshy clearing and pause to watch two male puku sparring across the river.

Suddenly the long grass between us and the river moves and warns us of something big, bold, and bulky.

"Buffalo!" Mark whispers, pulling me behind him and raising his rifle. A huge bull staggers out of a reed-filled dambo only fifteen yards ahead, his hooves sucking loudly in the mud. He pauses, lifts his head, and stares in our direction, nostrils flaring, a bunch of grass sticking out of his mouth.

"Freeze," Mark whispers to me, thumbing the .375 rifle's safety off. The bull lowers his head and starts toward us like a Mack truck, his horns swinging from side to side. He is looking directly at us, but — incredibly — does not seem to see or smell us. At only ten yards away, he stops again and raises his massive head to scrutinize us, blowing puffs of air as he tries to take our scent. Lone bulls can be extremely aggressive and often attack with no apparent provocation. Turning my head carefully, I see a tall winter-thorn acacia tree forty yards behind me.

"Mark," I whisper between clenched teeth. "I'm going to run to that tree."

"No! Don't move!"

The buffalo raises and lowers his head, as if straining to make out the two forms standing in front of him. It is said that buffalo cannot see very well, but he can't possibly miss us at this distance. Holding his head at full height, the buffalo starts toward us again. This close, he is going to be very unhappy when he discovers us. I take a tiny step backward.

"FREEZE!" Mark hisses. The bull stops, shaking his head.

"I AM RUNNING TO THAT TREE!" I hiss back.

Without moving his lips, Mark vows, "If you move, I'm going to shoot you in the back!" I pause, considering my options.

Eight yards away the bull is raising and lowering his heavy black boss, and again blowing air through his wide nostrils. It's too late to run. I couldn't make it to the tree anyway, with my knees shaking like this. Slinging saliva and grass stems, the buffalo shakes his head and grunts. He stomps and rakes his right hoof. A dank, musky odor lies heavy on the air. Lowering his nose to the grass, he smells along the ground and begins walking again in our direction. After a few more steps he looks directly at me, drool falling from his mouth. Once again he tosses his head.

He turns slightly, then step by step walks right past us until he enters the tall reeds on the other side of the clearing.

When he is out of sight, Mark turns to me and grins. "See, it's okay."

Sitting down heavily on a log, I ask, "Would you really have shot me in the back?"

Mark smiles and sits next to me. "You'll never know."

○ ○ ○

On our last day in Hidden Valley, we hike with Kasokola, Mwamba, and Simbeye up into the mountains. Believing that we are the first people to explore this little corner of earth keeps us cheery and warm in spite of the constant drizzle.

But then we come upon a well-used footpath leading down from the scarp mountains to the valley. "Poachers," Mwamba says. Walking silently, we follow its switchbacks across the hills. Periodically there are clearings and old campfires, and occasionally large meat racks and piles of discarded bones. The path has been used regularly for years by large bands of commercial poachers.

We tell ourselves not to be discouraged; after all, we know that poachers hunt in North Luangwa. Even so, our hike to Hidden Valley has taught us that this is a very special place, and even more worth saving than we had believed before.

o o o

The stall warning blares constantly as the plane shudders and shakes, trying to maintain altitude. We are flying our first wildlife census of North Luangwa, which requires that we hold an airspeed of eighty knots at two hundred feet above the ground, along transect lines running approximately east and west across the park. Because of crosswinds and thermal air currents, it isn't easy to hold a precise heading, airspeed, and height above the ground, even over the relatively flat floor of the river valley. We have divided the park into sixty-five grids, and as we fly along we use tape recorders and a stopwatch to record the species of wildlife, the habitat type, and the exact time, which will later be computed against our airspeed to give us a rough grid position for the animals sighted. From these data we will calculate the distribution and density of the animal populations.

But the mountains of the Muchinga Escarpment cut a jagged line across the park, jutting more than three thousand feet from the valley floor in a series of ever steeper hills, deeper gorges, and higher cliffs. No matter how rugged the terrain, for purposes of the survey we must maintain a constant height above the ground. When we reach the first foothills, Mark pulls up the plane's nose and pushes in the throttle to climb over the first range of hills. But as soon as we have crested them, he hauls back on the throttle, drops flaps, and we sink into the next narrow valley. We level off, and before our stomachs can come down, a rocky precipice stares us in the face and we are climbing again. At the top of the next ridge, the earth plunges away to a seemingly bottomless ravine, then soars again to new heights. Mark pushes the stick forward and the plane dives through the gorge toward what looks like the center of the earth.

Under these conditions we cannot possibly maintain two hundred feet, and at times the wings slice within feet of massive trees and boulders. I try to concentrate: "5 buffalo, 10:25, brachystegia woodland; 3 wild pigs, 10:44, upland mopane." But more than once I close my eyes as the plane tries to scale a rocky cliff. The mountains stretch for more than thirty miles across the park and we must fly across them thirty-two times. The stall warning sounds continuously and the plane barely maintains altitude. I glance at Mark. Beads of sweat glisten on his forehead, and his hands strangle the stick as he fights the downdrafts and swirling air currents. I think we must have reached the top; but as I dare to look up, the tallest peak yet fills the plane's windshield.

Mark stomps left rudder and turns the plane away from the mountain. Gliding at a safe height above the ridges he flies us back to camp and lands on the airstrip. I step out on wobbly knees and control an urge to kiss the ground.

"Okay, that's just not safe," Mark says. In all our years together, I have never heard him say these words. I lean against the fuselage while he spreads the maps across the plane's tail. "This is what we'll do," he says. "Instead of flying east and west across the scarp, we'll fly northeast and southwest along it, and sample it independently of the valley floor. The peaks and valleys won't be as severe that way. As long as we design the transect lines correctly, we'll cover the same ground and not distort the data."

After redrawing the lines and calculating the new headings, we take off again. True, the ups and downs above the jagged earth are not as severe flying along the scarp, but the plane still struggles to fly and I still struggle to keep my eyes open. But "scarping," as we have come to call it, is worth it. The soaring forested peaks surround not only fall-away canyons but soft grassy glades and mountain streams. We see a herd of sable galloping through a meadow and a family of elephants walking on an ancient path through the hills. A leopard balances on top of a termite mound as he watches us pass. With each transect line, each hilltop and dale, we learn more about Luangwa.

○ ○ ○

In late February there is an unusual break in the rains. The ground dries up a bit, and we are determined to cut a track across the park to the plains that parallel the Luangwa for miles. From the air we have seen more wildlife on these seemingly endless savannas during the rainy season than anywhere else in Africa.

We fly from camp on the Lubonga River, past Mvumvwe Hill and all the way to the Luangwa, searching for the route with the easiest stream crossings and the lowest hills. Once we have decided on a general route, Kasokola, Mwamba, Simbeye, and I drive out and cut our way across the bush the best we can. Periodically Mark flies overhead and gives us detailed instructions by radio.

"Delia, you're heading too far north, you're going to run into a gully. Pull back to the last streambed, then head zero five five degrees for about four and a half miles till you get to a rocky cliff. I'll tell you where to go from there."

"Okay, Roger. I copy that." And on it goes for two weeks, until the guys and I finally reach the Fitwa River. It is too deep for the Cruiser to cross, so we turn back for camp to collect Mark and the Mog. We pack the truck with darting gear, camping gear, plant presses, and food in preparation for a long expedition to the plains to dart lions, take wildlife censuses, and sample the vegetation.

"Mark, do you really think the Mog can make it through this?" I stare at the angry Fitwa, raging between mud-slick banks that are fifteen feet high. We will have to ease the Mog down a fifty-degree slope of mud, ford the shoulder-deep current, then climb the opposite bank.

"No problem. Hold on, guys," Mark calls to Simbeye, Kasokola, and Mwamba, who are perched on top of the gear in the back. Mark shifts to the third of the truck's sixteen gears and eases forward over the edge. At first the Mog's four-foot-tall mud tires hold on the greasy slope, their ribbed tractor lugs biting deep. Then, as the entire weight of the truck heads downhill, the treads break loose and we begin a sickening slide toward the river. I grab for the handles in the cab and hold on. The truck hits the river, submerging its front end completely under water, spray, and mud. Mark rams the truck into a higher gear. It churns through the river, water boiling through its chassis, around its sides, and into the cab around our feet. Near the opposite bank he guns the turbo-diesel and the Mog claws its way up the slope — so steep that we are almost lying on our backs in our seats. The big tires spin, slide, and sling mud through the window into my lap.

"Hey! Hey!" Shouts come from somewhere behind us. Turning, I see our three helpers tumble off the back of the truck and into the river, followed by our mattresses, tents, and bedrolls. Splashing about in confusion, they grab overhanging branches and hang on against the swift current. With a free hand or foot they snare bits and pieces of our gear as they float by. I jump from the cab and slide through the mud to help them, as Mark maneuvers the truck to the top of the bank. Pulling Mwamba from the current I say, "Thank goodness you guys can swim!"

"I can't," he says with his ever-present smile. "But it would be a good thing to know how to swim when crossing a river with this boss."

Unbelievably, nothing is lost. We repack the soggy gear and carry on. Two days later we reach the plains.

As we emerge from the woodlands, the savannas stretch for miles in every imaginable shade of green. Dozens of different kinds of grasses wave with iridescent reds, greens, and yellows. Like a huge abstract watercolor, they boast an array of spiraling blossoms and sprays of soft seeds. Choreographed by the breeze, they bow, swirl, and pirouette. Wind and grass make perfect dance partners.

And it is here where most of the wildlife of Luangwa spend the rainy season. Large herds of zebras, wildebeests, eland, impalas, and puku graze the succulent wild rice (*Echinocloa*) and other grasses (*Erograstis, Spirobolis*) and attract lions, leopards, spotted hyenas, and wild dogs. In the late afternoon they pause to drink at hundreds of small water holes filled with knob-billed ducks, black-smith plovers, jicanas, sperwing and Egyptian geese, and Goliath herons. Majestic crowned cranes strut nearby. Some of the smaller water holes are more mud than water, and in each a battle-scarred old bull buffalo wallows, caked in gray sludge, oxpeckers sitting on his broad back. Herds of more than a thousand younger bulls, cows, and calves mow the grass all around. Warthogs kneel to root up bulbs and subterranean delicacies. Greater kudu stand silently in the shadows of *Croton* and *Combretum* trees, and families of elephants pull up bunches of the tender grasses.

Amazingly, some of these "plains" are the same devastated and degraded woodlands we saw in the dry season. Then they looked like a portrait of ecological disaster — the soil a gray, dry powder without a single living plant to its credit, the mopane trees limbless and dead, the bushes leafless. We worried that this ruined habitat could never recover, but the rains have transformed it into a lush grassland that, in this season, is the most heavily utilized habitat in the park.

Most of this area was once healthy mopane woodland. But about fifteen years ago, when commercial poachers invaded the area and began killing unprecedented numbers of elephants, the harried survivors sought sanctuary deep in the heart of the park. Squeezed into these woodlands in unusually high numbers, the animals stripped the trees of their bark, leaving them vulnerable to diseases such as heart rot, and to the wildfires set by the poachers each dry season. This combination of pressures from elephants, disease, and fire has killed off hundreds of square miles of mopane forests, opening the way for the establishment of annual grasslands that appear only during the rains. The poachers have devastated the woodlands, just as they are devastating the elephants.

Now that the grasses have supplanted the dying woodland, there might seem to be no problem. But while grasslands favor buffalo and other grazers, forests are important to elephants, kudu, and other browsers. So poachers are reconfiguring the floral and faunal communities of the park. Who knows what changes in species composition or loss will occur if more woodlands are damaged by the bushfires that now sweep over 80 percent of the park every year.

○ ○ ○

Our radio tracking of lions in the Kalahari taught us not only about their natural history but about the habits of other carnivores, and the distribution and habits of their prey. We are anxious to put transmitters on the Luangwa lions so that we can learn more about their competitors and prey, and so that we can compare them with the Kalahari lions.

On the edge of one of the plains is a grove of large *Combretum obovatum* bushes that stand fifteen to twenty feet tall. The thorny branches of each bush hang down in a dense tumble, creating a spacious cavern within — a perfect place to hide. We put our pup-tent inside one of the combretums, and the guys erect theirs in another. In a clearing surrounded by a dense thicket, we build a campfire and set up the small table and chairs. With the Mog hidden in the bushes behind the camp, we are totally concealed from the wildlife on the plains.

We have always darted carnivores from a truck, but since the Mog is so huge, Mark worries that lions will not come within range of the dart gun. Instead, we will make a blind and operate from it. We hang a large piece of awning cloth between two combretum bushes not far from our little camp and hide our chairs, tables, and darting gear behind it. It is dusk by the time we finish, so we retire for the night in our tucked-away campsite.

At dawn we quietly cook a breakfast of oatmeal and fried toast over the campfire. Our plan is to set up a huge stereo speaker at each side of the nearby blind and play recordings of lions feeding, mating, and roaring their territorial challenge, in the hope that these sounds will attract a resident lion for darting. While I am making coffee, Mark plays the roars in the middle of our camp's little kitchen.

"It'll take a while for the lions to come," he says. "So we'll just finish our coffee, go to the blind, and play more roars there."

Meanwhile the speakers blare their great roars across the plains.

Quickly finishing their breakfast, Kasokola, Mwamba, and Sim-beye take refuge in the Mog, where they wait for us to call them. Mark and I stand together, sipping our steaming camp coffee as we watch the herds of zebra and wildebeests on the plain.

"Hey! There's a lion!" Mark exclaims, pointing straight ahead. A large male with a full mane stands less than sixty yards away, his chin raised, moving his head side to side, apparently searching for the intruder who is bellowing in his territory.

"Hurry, let's get to the blind," Mark says, as he picks up a speaker and sneaks through the grass toward the blind. I lift the other speaker and follow. Without the taped roars to home on, the lion heads off in the wrong direction and disappears from sight.

In the blind we quickly hook up the stereo system and start playing the roars, taking cover behind the cloth. Almost immediately the lion reappears, trotting straight for us. Careful to stay below the level of the awning cloth, Mark loads the darting rifle with a drug-filled syringe. At fifty yards the lion slows to a saunter, his eyes wide, tail flicking. It only now occurs to me that nothing but a piece of cloth stands between us and him. The tape player continues to challenge the male, who hunches his shoulders as he stalks toward us, stiff-legged. He breaks into a trot again, his mane swinging loosely.

Mark turns off the speakers and takes aim with the darting rifle. Since the lion is coming straight for us, the only possible target is his forehead. Mark holds his fire. At forty yards the lion veers left and circles the blind, just out of range, then disappears behind our thicket. The brush behind us is so thick we cannot see him. Mark plays the roars again. We stand in the silence, watching and listening.

"He's right here," Mark whispers.

"Where?"

"Right *here!*" Moving very slowly Mark points to the end of the blind where it is tied to the bush. The lion's muzzle is almost touching the cloth as he peers into the depths of the thicket — and our blind. He is at most a yard from my right leg; too close to dart.

He looks up and down the strange material, sniffing loudly at its unfamiliar odors.

Apparently satisfied that no other male lions are hiding in the bushes, he turns and walks toward the plain. Slowly Mark lifts the rifle above the blind and darts him in the flank. The lion whirls around and stares at us. We freeze again. This is the most dangerous moment: if he associates us with the sting in his rump, he may charge. He spit-growls once, but turns and trots away. Five minutes later he sits on his haunches, relaxing as the tranquilizer takes effect. After another five minutes he slumps to the ground, immobile. With the guys' help, we collar, weigh, measure, and ear tag the lion.

For the next three days we set up the speakers at different spots on the plain, and try to call in other lions. But wherever we go, it is this same male who appears, still trying to chase imaginary intruders from his area. We name him Bouncer.

o o o

We have just returned to Marula-Puku from an expedition late one afternoon, and the men are mending a flat tire on the Mog. We have long ago given them the hut to live in; Mark and I sleep in a puptent near a small supply tent at the other end of the island. Our kitchen is merely a clearing under the marula trees, with crates and tin boxes arranged in a square. I am mixing cornbread batter and preparing to bake it in the black pot by the fire.

"KA-POW! KA-POW!" Gunshots from across the river.

"Poachers!" Simbeye shouts, pointing south.

More shots. Three, four, five, six, seven. I feel their concussion in my chest. Eight. Nine.

"Sons of bitches! AK-47s!" Mark swears. "Not more than a half-mile from here. They're probably shooting elephants."

All my doubts about our fighting the poachers dissolve in an instant. "We've got to do something!"

But with only one gun and no authority to go after poachers on our own, all we can do is drive to Mano to get the scouts. Their camp is eighteen miles northwest of Marula-Puku. Within minutes of the last shot Mark and I are in the Mog, clawing our way up the scarp. When we arrive, Mano camp is quiet and still, although scattered cooking fires glow here and there in front of each hut. Mark jumps down and quickly tells Island Zulu about the gunshots. A few scouts stand around, leaning against the truck. Tapa, who had dried fish and meat in our camp, yawns. Nelson Mumba, still wearing his red bandana, walks away. As honorary game rangers and directors of our project, Mark and I have the authority to order the scouts on patrol. But we do not want to command them; we want them to come on their own.

"We have no ammunition," Zulu tells us.

"What happened to it?" Mark asks. Mosi Salama, the warden in Mpika, swore to us that each man had been given his monthly allotment of five rounds. The scouts look at each other, speaking in Chibemba. As before, all agree that they have not received their allotment.

"I have one round," says Gaston Phiri, a lively, short man with the energy of a shrew. "But we have only four rifles, and one does not have a bolt."

Mark senses a faint willingness in Phiri. "Mr. Phiri, I will pay every man who comes with us two hundred kwachas for each poacher he catches."

"But we have no food for patrol," says Phiri.

"We will give you food," I interject.

Eventually six of them agree to come, but they will need two hours to get packed. We urge them to hurry, so that we can catch the poachers before daybreak, but Phiri tells us, "We cannot patrol at night. That is when the lions are hunting. Don't worry. The poachers will not move at night either."

It is midnight before we get back to Marula-Puku, and Mark and I are up again at four-thirty, getting a big fire going, making as much noise as possible to wake the slumbering scouts. They finally join us around the fire at five-fifteen. They unpack their worn knapsacks, which they have just packed, and Phiri puts a huge pan of water on the fire to boil their n'shima. The others lie around the fire smoking tobacco and chatting, waiting for the water to boil. I stoke the fire continuously, to make it hotter.

"Look, it's boiling," I say to Phiri.

"Yes, but it must boil for many minutes to get very hot."

I stare at him, wondering how I can explain that water cannot get hotter than it is at the boiling point. The poachers are probably breaking camp and moving on, and here we are waiting for water to perform a miracle. But I keep quiet, hoping that the poachers are also waiting for their water to exceed the boiling point.

Twenty minutes later the water is hot enough to satisfy the scouts. They toss in handfuls of mealie-meal, stirring vigorously, then eat the paste with the bully beef and tea we have provided. They breakfast in leisurely fashion, as is their custom; when they are finished, they wash their hands carefully in a basin of water. Then each man rolls a cigarette and smokes it down to his fingers. An hour and a half after rising, they are nearly ready to head out.

For patrol, we give them enough mealie-meal, salt, dried fish, beans, tea, and sugar to last a week. They repack their bags and at six forty-five announce they are ready. Mark steps forward to go with them, but Phiri holds up his hand.

"You can tell us, please, where you heard the gunshots. But you cannot escort us there. You are not a scout."

Mark is annoyed but accepts their position. He takes the compass and, pointing 170 degrees, shows them exactly where we heard the shots, and adds, "Don't forget. For every poacher you capture, you each get two hundred kwachas."

We wish them luck as they march out of camp in a ragged line. Four have uniforms, two do not. Their trousers are torn and patched. Four have boots, one wears a pair of rubber-tire sandals, and one is barefoot. Only two have proper rucksacks; another carries a plastic bag, and Phiri has an old gunnysack thrown over his shoulder. Island Zulu is still lugging his old cot frame and has three plastic mugs, red, yellow, and blue, tied to his pack. They have three rifles — one without a bolt, another without sights — and a shotgun that won't extract spent rounds. And they have only one round of ammunition. No wonder they are reluctant to go after poachers armed with AK-47s. Nevertheless, they are going. "Good luck," I say again softly.

Mark, Simbeye, and Mwamba go to the airstrip to grade it with the Mog's loader, while Kasokola and I remain in Marula-Puku.

Kasokola, the youngest of the Bembas, is shy and quiet, but smiles readily at the slightest prompting. He and I clear the grass where the office, bedroom, and kitchen huts will go, then drive sticks into the ground to mark the corners of each. After seeing what happens to a mud hut in the rains, we plan to build stone cottages with thatched roofs.

At eight-thirty I see the first vultures circling to the south, where the shots were fired. The carcasses are closer than we thought. The game guards should have found them easily by now and be on the trail of the poachers. I feel sickened that elephants were shot so close to camp, for if we can't protect the animals in areas right around us, what chance do we have to stop poaching in the rest of the park?

"They are coming, Madam," Kasokola says quietly.

I whirl around, "Who is coming?"

He points south. "The game guards."

The scouts are walking toward us, spread out in a line along the riverbank. It is eleven o'clock. Have they captured poachers already? I hurry toward them, as they emerge from the long grass. There are only six men; no poachers.

"Good morning, Madam," Gaston Phiri says loudly and cheerfully.

"Hello, Phiri," I answer quickly. "What is happening? Where are the poachers?"

Phiri raises himself to his greatest height, which is about the same as mine, and announces, "We have found two poached elephants! They are on the riverbank, only half a mile from here."

"Good," I say, "we sort of knew that." I point to the scores of vultures soaring less than a mile away. "But what about the poachers?"

Still holding himself erect, Phiri answers, "We thought you would want to take photographs of us with the dead elephants."

I stare at him in disbelief. "Phiri, we have plenty of photos of dead elephants. In fact, since we have been in this national park, we have only taken pictures of dead elephants! Since we have been in Zambia we have seen more dead elephants than living elephants. What I want is a picture of you with poachers. You are supposed to capture the poachers, not have a photo session with the dead elephants."

The scouts frown and look away, clearly disappointed that they are not going to be photographed.

"We have found the tracks of the poachers," Phiri continues. "We can chase them out of the park."

"Phiri, they will be miles from here by now. They probably left at sunup."

"We will go after them," he says, "but first we must have lunch. Can you give us some more canned meat?"

"We gave you a week's supply of food a few hours ago. We have only four tins left. Okay, here are two of them. But please try to capture those poachers!"

"Thank you very much! And anyway, we thought maybe you would like to take our photograph here before we go on patrol."

"Right. You're absolutely right. We should have done that this morning." I don't even bother to rush as I bring the camera from the tent. "Stand together over here."

They crowd together, holding up their rifles and making the most fearsome facial expressions and guttural growls. Too bad the poachers have not joined us for lunch — they might have had the fright of their lives. Little do the scouts realize it is a picture of despair.

The scouts do not come back to Marula-Puku. Later I ask Simbeye if he has heard anything, and he tells me that when the scouts left our camp they marched straight back to Mano to share the food with their families. Mark and I are disappointed, but realize we may be expecting too much of six men with one bullet and hungry families. We will have to equip them better, not only with rifles but also with tents and uniforms, and somehow ensure a more reliable supply of food at Mano. By candlelight we draw up a list of the scouts' most pressing needs and resolve to discuss it with the warden on our next trip to Mpika.

Two days later we see the elephants of Camp Group moving in the distance along Khya Stream. Now, instead of ten, there are only eight.

8

The Heart of the Village

DELIA

They learn in suffering what they teach in song.
— PERCY BYSSHE SHELLEY

o o o

THE SHARP EDGE of the hand-carved wooden stool cuts into my thigh, as I glance back and forth from Mark to the village headman, who is acting as our translator for Chief Mukungule. We are all sitting on various-sized stools in the thatched n'saka — except for the chief, who is enthroned on a seat taken from an old DC-10 aircraft. The chief's barefoot wife squats on the earthen floor near a clay pot of steaming sweet potatoes. Crowded nearby, their heads leaning toward us to catch the words of the translator, are twenty to thirty villagers of Mukungule.

"You see," Mark says, "if we can save the wild animals, the tourists will come from America to visit the park. They will bring money that will benefit your village. And if we can help bring back the animals around the park, your people can hunt them for meat — in a controlled way — so that there will always be some for food."

The headman stares at Mark for a long moment before turning to the chief and delivering his translation in Chibemba. Mukungule was born in 1910 and has been chief since 1928. He claims to have one hundred eighty grandchildren. His once-dark Bantu eyes have faded to a piercing ice-blue, but his personality is as warm and gentle as the African breeze. Most of his teeth are missing, no doubt the result of his insatiable appetite for sugar. He knows no English, but he smiles easily and communicates well with those ancient eyes, which have seen many things, including the coming and going of the British.

After an exchange in Chibemba, the headman turns to Mark, "We do not know this word 'tuureest.' "

I decide to give it a try. "Many people in the world like to travel, and they take money with them to pay for food and places to sleep. So the country they are visiting gains money. There are no elephants or Cape buffalo in America or Europe. The people who live there will come here to see these wild animals. They will pay a lot of dollars and pounds to see them. Your village could benefit from these visitors."

As the headman translates, the chief looks at us and nods his head slowly.

"But if the poachers keep shooting the elephants and buffalo, there won't be anything for the tourists to see. They won't come here. They won't bring their money."

I watch the villagers' faces; only the chief is nodding. Most of the people of Mukungule are of the Bisa tribe, which split from the Bemba tribe after the great exodus from Lubaland in Zaire in the nineteenth century. The older villagers, including the chief himself, remember well the days when they lived in the valley at the Lubonga-Mwaleshi confluence, the very spot inside the national park that we have come to love. The chief's ancestors, generations of chiefs, are buried under a tree only a few hills from our camp. Every October the headmen trek down the scarp to pay their respects to those spirits, by tying white swatches of cloth in the branches of the tree. It was their land, their home, their hunting ground, their burial place. As far as anyone knows, they were the first humans to live there — and until we came, the last. When the area was designated a game reserve by the British colonial government in the 1950s, the Bisas were asked to leave. Now we live near their ancestors and ask them to stop shooting the animals for meat and the elephants for ivory. We talk of "tuureests" coming from the moon with a kind of money they cannot imagine.

They left the valley willingly, the chief tells us, because the Mwaleshi hippos ate their crops and made farming impossible. In those days there was plenty of wildlife on the scarp, right here in Mukungule. As long as the Bisas hunted with bows and arrows, they could not kill too many animals and there were always enough to feed the tribe. Once guns were introduced, hunting was easier and no one controlled the numbers of animals shot. Soon all the animals west of the park were gone.

There is no butcher shop in this entire area, the chief continues, not even in Mpika. If a man in Mukungule wants to put meat on the table for his children, he must poach in North Luangwa National Park. I shift uneasily. Unfortunately there are so many people needing to feed their families, and so few animals left, that allowing people to kill "for the pot" would amount to a quick fix that would soon eliminate all wildlife in and around the park. For this reason, subsistence poaching — killing to put meat on the table — must be controlled as well as poaching for commercial gain. Both will lead to the near-term destruction of a valuable resource that can be used to raise the living standard of the people.

Talking a bit faster now, we explain that we understand their problems: there are no jobs in Mukungule, no meat, little protein. We want to assist them. If they will stop poaching, we will help them find other jobs and other sources of protein. Later, if tourists come to the park, there will be lots of work for the men; and the women can grow beans and groundnuts, raise chickens, build fish farms, and sell produce to the tour operators. While obtaining our permits to operate in the park, we persuaded the Zambian government to return 50 percent of its revenue from tourism to the local villages. But if the poaching continues, most of the animals in North Luangwa will be gone in five years. Then what will they do?

"You don't have to wait for the tourists to come," we tell them. "Anyone who will exchange poaching for a job can talk to us right now." We have raised funds for this purpose in the United States through our new foundation, the Owens Foundation for Wildlife Conservation.

As we rise to leave, bowing and clapping our respects to the chief in the Bisa and Bemba fashion, most of the villagers drift away, apparently uninterested in our offers. Only a small group of about ten women, standing to one side and not smiling, beckon to us. They are dressed in tattered Western-style blouses with faded chitenges wrapped around their waists. All are barefoot, although one is holding an old pair of high-heeled shoes, apparently to show that she does own some. With the help of the translator, and after much arguing on their part and confusion on ours, we agree to pay the women five kwachas for every bundle of thatching grass they cut and groom. We will need about two thousand bundles, which would bring them a total of ten thousand kwachas, probably more money than the entire village has ever seen. After making a circle of wire to show them the requisite size of the bundles, we arrange to return for the grass in seven days.

o o o

"KLOCK, KLOCK," the stones crack loudly against one another as we drop them onto the pile. One by one, we collect thousands of them — from the rock bars and the hillsides — to build our camp. For over a year we have lived in our small puptent, now tattered by the wind, and we long to unpack our trunks, put our books on a shelf, and not have to crawl out of bed every morning like caterpillars. We make bricks from sand and sun on the beach of our little river, and lay the foundations. Slowly, bit by bit, stone by stone, the cottages rise from the ground.

We level the last of the termite mounds from the airstrip and a charming representative of the Division of Civil Aviation, Arthur Makawa, travels from Lusaka to certify it. Once he has done this, he makes a mount for our wind sock and even helps collect stones for our camp. Finally the stone walls are complete, ready for the thatch roofs.

A week after our trip to Mukungule, Simbeye, Mark, and I drive back up the scarp, pulling the trailer to collect the thatching grass. Mukungule consists of widely scattered family bomas, surrounded by their fields. One cluster of mud huts is perched on a hilltop, another is nestled in a banana grove near the stream, and in the distance another is situated in a grassy field. In the center is a mud-brick, one-room courthouse and a three-room schoolhouse. We park next to a large family boma where five mud huts surround a cooking fire. Several small piles of grass bundles are stacked along the road. We call a greeting to two women, sitting at a fire, who wander over and speak to Simbeye in Chibemba.

"This is all they have collected," he tells us. "There were many problems. Some of the women snitched the cut grass from their neighbors and added it to their own piles. Then some villagers stole the grass for their own houses. The women became discouraged and quit working."

Even the bundles stacked by the road are less than half of the size agreed upon.

I look around. Villagers are standing by their huts and watching us. A few women have their heads together, giggling. What am I doing here, standing in this field of weeds talking about grass? I'm supposed to be studying lions, thinking about kin selection, calculating degrees of variance. How could I ever explain to our colleagues, who are waiting patiently for our scientific papers on brown hyena social behavior, that I am negotiating — unsuccessfully — the cutting of grass! I hug my arms to my chest and look to the sky.

"I miss Africa so much, and I am standing right in the middle of it."

"Let's get out of here," Mark says. "We'll try another village."

o o o

Tucked among the green folds of the escarpment hills, deep in the miombo woodlands, lies the small village of Chishala. It is closer to the national park boundary than any other village, and notorious for its poaching. We have been told that almost every adult male in the area has an illegal gun and hunts in North Luangwa. The pole bridge that once linked Chishala to the outside world was swept away by a nameless storm some years ago, so that now the road is only a footpath winding through the forest.

On a Tuesday morning we ease the Land Cruiser down the path, ford the river, and drive into the center of the village. In a shady spot near a crumbling mud hut, a couple of dozen men are sitting on the ground around a central pot frothing with homemade beer. Immediately, the men stop talking and stare at us. We have not been to this village before, but since there are no other white people in the area, they must know who we are. It is obvious from their cold glares and silence that we are not welcome.

I am tempted to turn around and leave, but Mark steps out of the truck, so I follow. Two of the men get up and walk quickly into the forest. The others continue to stare. One says something in Chibemba and they all laugh. When we are twenty yards from them, Mark calls a friendly greeting. No one responds. Suddenly an old woman rushes toward us, mumbling excitedly, and reaches out for Mark's arm. In the tense atmosphere her movement startles me, and I whirl around. She shrinks back, bowing and clapping in submission. Her sagging face, neck, and breasts are etched with a thousand fine wrinkles. Her deepset eyes are watery, and the hand that she still reaches out to us is knobbed and gnarled from the toil of primitive life. Pleading with us in Chibemba, she points to a cluster of banana trees where a young woman sits, holding a small, limp child.

Mark motions to me to get the first-aid box from the truck, then we both bow to the young woman and kneel beside her. The little girl in her arms appears to be two or three years old and is slumped against her mother's chest. Her thin arms and legs dangle from the soiled rags wrapped around her body. Her eyes are closed, but her tongue works constantly against her dry, cracked lips. When Mark carefully moves her head, it drops heavily to one side.

"Diarrhea?" he asks the mother. She nods and whispers in English, "Bad diarrhea, four days."

Mark asks the old woman to wash out an enamel cup, while I dash back to the truck to get clean drinking water. Mark stirs a rehydration mixture from a packet into the clean water and hands it to the mother. "She must drink this slowly." Holding the child's head up, the mother touches the cup to the girl's mouth and encourages her to sip the liquid. The tiny lips move and after a few swallows, large brown eyes open wide and stare into ours. Grasping the cup with both hands, the girl tries to drink faster, and Mark motions to the mother to go gently. I steal a glance at the beer circle. The men are watching us closely.

Rehydration, although not a cure for diarrhea, can produce a miraculous recovery from the symptoms. By the time she has drained the cup, the little girl is holding her head up and looking around at the strange white people and their big truck.

"Feeling better?" Mark asks. She buries her head in her mother's neck for a moment, then turns again to look at us. Her grandmother grins, claps, and bows to us, and the mother smiles shyly. The men in the beer circle are talking quietly to one another. One of them — the father — stands and walks over to the banana trees, squatting next to the mother and child.

Speaking in slow, short phrases, Mark explains to the parents that their daughter is still very sick and should be taken to the hospital. But the mother says she has no way of getting to the nearest clinic, which is in Mpika, sixty miles away. If we do not treat the girl ourselves, she could die in a few days. Mark takes some Bactrim from the first-aid box, cuts the tablets into quarters, and explains the dosage to the parents. We also give them more rehydration mixture. The mother bows her head and whispers, "Natotela — thank you." The father lowers his head and nods.

We gather our gear and walk boldly to the beer circle, greeting the men in Chibemba. Most of them smile and nod. Speaking through a translator, Mark and I tell them about our project and how we want to help them find other jobs so that they will not be dependent on poaching. The men laugh wildly at this. We tell them we have not come to arrest anyone. No matter how much poaching they have done in the past, it will be forgotten if they lay down their weapons and take up a new job.

"How soon?" one of them asks.

"Right now, if you want," Mark says.

Within thirty minutes, everything is arranged. The women will cut thatching grass and carry it on their heads to the grooming station. The young men and children will groom the grass on large combs, which they will build with lumber and nails that we will provide. Apparently the adult men will supervise. One old man, who calls himself Jealous Mvula, volunteers for the job of night watchman. We agree on a bundle size and a price, and when they calculate how much money they can make, a ripple of excitement passes around their faces. Telling them that we will be back in two weeks, we climb into the truck and drive away. I see the little girl, still in her mother's arms, waving good-bye to us.

Upon our return in two weeks we discover a veritable grass

factory. Swaying dramatically to balance the large bundles on their heads, women file through the fields toward the grooming station. Ten large combs have been constructed; young men and older children pull grass through the teeth to separate the leaves from the stems. Huge piles of grass, the size of haystacks, surround the decaying huts. Young men sing as they bundle the groomed thatch and throw it onto the heaps. The adult men are still sitting around the beer pot, but now and then they call orders to the others.

As we step out of the truck, the child we had treated runs up to us, smiling. Her mother, laboring under a large grass bundle, waves from a distance. Greeting everyone in Chibemba, we tell them how pleased we are with their work. Jealous has kept a detailed record of how many bundles each villager has collected or combed, and we calculate how much money we owe each of them. As we pass out the kwacha notes, people nod to us solemnly. Jealous tells us the village has made more money in two weeks than in two years of poaching.

While Mark and the men begin the scratchy, cumbersome task of tying hundreds of bundles onto the trailer, I sit on the bare earth in the shade of the banana trees and call to the children to gather around me. Holding up *International Wildlife* and *Ranger Rick* magazines, I begin to explain the pictures. But I do not realize that these children have never seen a color photograph. As I hold up a glossy centerfold showing an elephant family grazing on a savanna, the children "oooh" and "aaaah," clasping their small hands over their open mouths. Squeezed tightly together in a semicircle, they lean forward, taking in every detail of the photo. One young boy reaches out a finger to touch the shiny page, as though he expects the elephants to be there. I feel a tightness in my throat and a tear in my eye.

They do not seem to know that real elephants move silently through the forest just beyond the scarp mountains, and that their own fathers have slaughtered them for years. When I ask how many of them have seen an elephant, they all shake their heads. They are five miles from North Luangwa National Park and have never seen a live elephant!

Finally the truck and trailer are loaded with grass. The headlights look like eyes peering out from under a floppy straw hat. Brushing grass seeds and stems from his shirt, Mark joins me briefly, then we gently extricate ourselves from the circle of children, promising to return with more pictures. We have so few magazines that we will have to take them with us to the next village. One little girl helps me pick up the magazines from the ground, holding them with both hands as though they are the most precious things she has ever touched.

As we turn to go, the men in the beer circle call to Mark to join them; he walks to the edge of the gathering and greets them in Chibemba. Some are sitting on stools handmade from stumps, others sit cross-legged on the ground around the large clay pot full of home-brewed beer. A single drinking straw made from a reed stands up in the thick brew. The men take turns bending over the straw and drinking. With a sweep of his hands, one man invites Mark to share the drink. Diseases — dysentery, AIDS, cholera, tuberculosis — are common in these remote villages, as in most areas of Zambia. Yet Mark hesitates only briefly before bowing his thanks and kneeling beside the pot. Smiling at the men, he quickly pulls out the straw and turns it upside down. The men are silent for a moment, and I fear he has insulted them, but then they break into applause. Mark takes a long pull on the straw and exclaims, "Cawama sana — very good beer!" One of the older men walks to the pot and, with a great flourish, pulls the straw out, flips it over, and has his turn. The men rock back and forth, laughing.

As Mark thanks them and walks away, the men leave the beer circle and gather around us. Many are young — twenty-five to thirty-five years old — articulate, strong, and intelligent. For the first time, they admit openly that they are poachers but explain that they have no other jobs. "We want to conserve," says one man named Edmond Sichanga, who is dressed in clean, Western clothes, "but there is no job in conserving." I am surprised to hear him use the word "conserve," and even more surprised to hear him sum up in so few words one of the most critical environmental problems.

"That is often true in the short term," I say. "But in the long term, there will be more jobs if we can conserve North Luangwa."

"We can hire fifteen of you right now to work on the road between here and Mukungule," Mark tells them. "Then you will have jobs and a new road to your village. There will be more jobs later."

"That will be very good. Also we have another problem," Sichanga says, but now his eyes are twinkling with mischief.

"What is that?" Mark asks.

"We do not have a soccer ball. You can see over there, we have made a football field, but we have no ball."

"Okay," Mark says, "you stop poaching and we'll bring you a soccer ball. Is that a deal?"

The men raise their arms in salute. They shake our hands over and over in the Bemba fashion. Is it really going to be this easy to stop poaching in North Luangwa? We give out a few jobs and soccer balls and save the elephants?

The entire village stands along the road, waving and cheering as we drive away in our hay wagon. We creep along at a sluggish pace, and I lean backward out of the window to be sure the load isn't shifting too badly.

We have gone only a few miles when an African man dressed in a tattered coat steps in front of the truck and hails us. It is Jealous, the night watchman. On our first visit to Chishala Mark told him that we would pay for information about the poachers. He stops the truck. Jealous rushes toward us, his coat fluttering behind him, and hops onto the running board. Sticking his head through the window, he says: "These men of Chishala, they are not the big hunters. They kill a few elephants, but the men who kill many, the big poachers, they stay in Mwamfushi Village."

"How do you know this?" Mark asks.

"I have two wives. One lives in Chishala, the other in Mwamfushi."

"Who are the big poachers? Do you know their names?" We have heard of Mwamfushi Village; the game guards were chased away by men armed with semiautomatic weapons.

"The big hunters are Simu Chimba, Chanda Seven, Bernard Mutondo, and Mpundu Katongo. But Chikilinti is the worst — he is the godfather of them all." Jealous jumps from the running board and disappears into the forest. I quickly write down the names, underlining Chikilinti twice.

o o o

Fording transparent rushing rivers, building rickety pole bridges, clearing overgrown tracks, and slogging through the mud, we visit the remote villages along the western border of North Luangwa, each one more difficult to reach than the last. At every village we meet first with the headmen and elders, and talk to them about what jobs and food are available, what skills the villagers have, what materials they need to start small industries. We promise to help each village set up at least one cottage industry such as a carpentry shop, a sewing club, a fish farm. We talk with the headmaster of each school and organize a conservation program that includes a wildlife club for the children. Hauling around a generator to provide electricity, we set up our projector and show slides of wildlife to the children and adults. Each school has a counterpart in the United States, which will later send art supplies and letters to the Zambians. Smoke Rise Primary School in Atlanta, for example, is paired with the primary schools in Mpika and Nabwalya.

Not always welcome, we are sometimes warned and threatened. Over and over we hear that the worst village for poaching is Mwamfushi, and that it is not safe for us to go there. We keep visiting new villages and returning to old ones, slowly recognizing faces and making friends. Mr. Chisombe at Katibunga wants to build a fish farm. Syriah of Chibansa, only thirteen, wants to raise rabbits and ducks. The people of Fulaza, who for years have traded poached meat from North Luangwa for ground maize, need a grinding mill.

Determined not to make the mistake of creating welfare villages (as aid organizations unfortunately have done for years), we give nothing away. Anyone we agree to help must promise to stop poaching. And once their businesses are going, they must repay the original loans to our project. Before we hand out any money, they must contribute as much as they are able to their new enterprise. If they want a grinding mill, they must build the mill house with mud and grass before we will buy the mill. Then we know that they are committed to the enterprise.

Mukungule, where the women failed to cut the grass, is in the center of the region. During the months of 1988, as we go from one village to another, we often pass by the chief's n'saka with its airplane seat. The word has spread that Katibunga has a fish farm and that Chishala made fifteen thousand kwachas from cutting grass. One day I am driving alone from Katibunga while Mark is flying antipoaching patrols in the valley. Mrs. Yambala, a teacher in Mukungule, stands in the road and waves for me to stop. She invites me to sit under the trees with her and tells me that the women of Mukungule want to start a sewing club. They can sell ladies' blouses, baby clothes, and tablecloths to the wives of the game guards, who have cash, she says. I tell her that we will buy the materials and once they are able to sell their products, they can pay us back. Her husband, headmaster of the school, asks me to start a wildlife club for the children, as we have done elsewhere. I shake hands and tell them that we are happy to work at last with the people of Mukungule.

o o o

"Okay, everybody," I call out, standing near the banana trees of Chishala, where the villagers cut our grass. "Let's all go to the soccer field. We have a surprise for you." Jealous and the school-teachers help me pass the message, sending runners to the distant fields and huts. I watch stooped and elderly villagers, vibrant young men, colorfully dressed mothers and children, move across the hills and through the meadow toward us. Soon an excited crowd has gathered at the field. I stand on the back of the truck, listening.

"Here he comes!" I shout. "See, the ndeke comes!" I point north to where our plane glides into view over the forested hills. The villagers wave and shout. Mark swoops in low, as though he is going to land on the field, and at the last second flings a soccer ball from the plane door. It falls among the young men and children, and a game begins on the spot. Everyone in the crowd, even an old man with a cane, takes a turn kicking the ball. After a while Sichanga, one of the men from the beer circle, picks it up, walks over, and thanks me. I can see that Mark has drawn an elephant on the ball and written, "Play Soccer, Don't Poach Elephants!" I point this out to Sichanga and his friends, who are now employed by our project to hand-grade the road from Chishala to Mukungule.

"Remember our deal," I say.

And they answer, smiling, that they remember.

o o o

The wide-open arms of the marula trees welcome us back to Marula-Puku after every village trip, when for days we have slept on the truck, eaten canned food, and bathed in cold streams. A roaring campfire, started by Simbeye, Mwamba, and Kasokola at the first sound of our truck, brightens the reed kitchen boma as we pull into camp. On the far side of the slow-moving Lubonga we can see puku and buffalo grazing. The guys greet us, no matter what time of day or night, and help unload the muddy truck.

The stone and thatch cottages, all but swallowed by the giant trees, are now complete. In a semicircle along the riverbank is an office cottage, a kitchen cottage, a bedroom cottage, and an open n'saka, the traditional Bemba meeting place. Solar panels power the few lights and two computers. At the back of camp is a workshop, stocked with tools and spare parts for the trucks. The camp is neat and efficient, yet its structures of local stone and grass blend into the riverbank so well that they are hardly noticeable from the far side.

One day as we pull into camp, Simbeye rushes up to us.

"Bosses, come, you must see," he shouts. As we climb down from the truck, he pulls on Mark's arm and leads us to the grassy area between the office and bedroom cottages.

"See, the tracks; he was right here." Simbeye points to the large footprints of an elephant only fifteen yards from our bedroom. "He has been coming here every night to eat the marula fruits. He is one of the Camp Group. There are eight, but only one comes here. The others feed on the hill."

"That's great, Simbeye! Did you actually see him? Isn't he afraid of you?" I ask.

"He comes only at night. He moves like a big shadow, so you cannot hear him. But I wait in the grass by my hut, and I see him come. He does not know I am there."

"Are you sure that it is the same elephant every night?" Mark asks.

"Yes, sir, I am sure. He is the one with tusks as long as your arm," Simbeye holds his hands about three feet apart, "and he has a small hole in his left ear."

After an early supper of cornbread and beans, Mark and I take up positions by the window in the bedroom cottage. Whispering in the darkness, we take turns peering into the night for large, moving shadows. But the elephant does not come. Finally, we fall asleep in our clothes on top of the bedcovers. In the morning there are no fresh tracks. The elephant must have known we were inside the cottage. After all, he is clever enough to have survived poachers' bullets for many years.

Early that morning, anxious to get back to the wildlife work, we drive to the airstrip to fly an antipoaching patrol. Suddenly Mark stops the truck. Eight bull elephants stand in a tight group only three hundred yards away, on the steep hillside overlooking the river. Wrapping their trunks around the bases of the tall grass, they pull up large clumps and munch on them. Not one looks in our direction. I hold my hands to my face, and Mark squeezes my shoulder. This may happen often in other parts of Africa — people watching a small group of elephants feeding — but never before have we been able to get so close. Instead of fleeing at the first sight of us, the elephants ignore us completely.

After that they show up regularly here and there. We see them from the plane near Hippo Pool, across the river feeding on marula fruits, late one afternoon in the valley beyond the airstrip. They keep their distance, but they do not run. Perhaps they have learned that they are safe near our camp.

The other elephants in the park are not so safe. Mark has been flying patrols daily, whenever he is not on a village trip with me. Every week he discovers four to six dead elephants. With each discovery we plead with the game guards to go on patrol, but there is always some reason why they cannot. They have not mounted a

single patrol on their own since we arrived last year. The radios
we ordered months before still have not been approved by the
government. Every time we want to get a message to the game
guards, we have to make the long drive up the scarp to Mano, or
else Mark has to fly over their camp and drop a message in a milk
tin.

A week after first seeing the elephants near camp, Mark takes
off on a flight to drop a message to the game guards, asking them
to patrol the hills around our camp to help protect the herd. No
sooner is he airborne than he flies into a flock of vultures. Hitting
one can be fatal, so Mark quickly banks the plane starboard to turn
away from the birds. Looking down, he sees the mutilated carcasses of three male elephants, sprawled in pools of blood and
splashed white with vulture dung. Swearing and flying dangerously
close to the treetops, Mark circles the area looking for the poachers. Seeing no sign of them, he lands and drives madly up the
scarp to collect the scouts. Four hours after spotting the elephants,
we stand with the guards around the carcasses — huge, gray monuments to a dying continent.

"Bastards!" Mark paces around. "They're laughing at us. You
know that, don't you?" Mark stares at Gaston Phiri. "They're
laughing at you! They know that they can poach right here and get
away with it."

"We know who shot these elephants," Phiri announces proudly.
"We can tell by their boot tracks in the sand. It is Chikilinti,
Chanda Seven, Mpundu Katongo, Bernard Mutondo, and Simu
Chimba from Mwamfushi Village."

"Well, good," Mark says. "If we know exactly who the poachers
are, we can go to the village and arrest them."

"Ah, but we cannot get these men," Phiri tells us. "They have
juju."

"They have what?"

"These men are real men, but they have a magic from Zaire.
They can make themselves invisible, so that we cannot see them.
They can stand right here among us, but we will look through
them. We ourselves can never capture them."

"Come on, Phiri! You don't believe that."

But Phiri insists, sounding hurt. "It is fact. It is like this: they stand under a tree, put on a special hat, pour magic potion over their heads, and turn in circles. Then they disappear."

"Phiri, don't you know it's impossible to be invisible?" Mark looks anxiously at the other scouts, hoping for support.

"Maybe for you, but not for these men. You may know what is written in your books. But these men, they know magic from Zaire!"

I drop my hands to my side and walk around in a small circle. Mark stands in silence, trying to control his anger, unsure of what to do.

"The men from Mwamfushi are the hunters," Phiri goes on, "but they use the men from Chishala as carriers."

"From Chishala!" I cry out. "Those men we gave jobs, and the soccer ball?"

Phiri just looks at me, and I get the message. How could we be so naive? Did we really believe we could win them over with a soccer ball and a few jobs?

"There is something else," Phiri says. "That man, the one who calls himself Jealous, the man who told you informations about the poachers. He was poisoned. His stomach is very sick, and his lips are burned very bad. He has been taken to the hospital in Mpika."

Mark lowers his head into his hands and stares at the ground. I walk away and look out over the golden, rocky hills. Something catches my eye. Five elephants — all that is left of Camp Group — move silently away through the trees. I make no sign that I have seen them, but I watch. As they reach the crest of the hill, I see the male with the small tusks — the one with the hole in his left ear. He has survived one more time. "Go well, Survivor," I whisper. "Go well."

9

Survivor's Seasons

DELIA

> Here you must look
> at each thing with the elephant eye:
> greeting it now for the first time,
> and bidding, forever, good-bye.
>
> — ANNIE DILLARD

o o o

SURVIVOR, FOLLOWED by the four young bulls of his group, trundles slowly up the small, rocky hill. It is May and the elephants are on their way from the plains to the great scarp mountains. Even though it is a short migration of only fifteen to twenty miles, they take several months to pass through the belt of hills along the base of the escarpment. The tall, waving grasses and small trees and scrubs (*Terminalia, Colophospermum mopane, Combretum*) make good forage at the end of the rains, while the rushing rivers and hidden lagoons provide water. But one of the main attractions of the area is the large, spreading marula trees (*Sclerocarya caffra*) that drop their sweet fruits at this time of year.

The elephants know where the marulas are. Well-worn paths lead from one to the other, and under each the grass is matted down where the large beasts have fed for hours on the yellow fruits. They walk to one of the marula groves near Khya Stream. Swaying gently back and forth, they feel and sniff along the ground with their trunks, then pluck the fruits into their mouths with loud slurping noises.

The five animals are in their twenties and form a loose-knit group of independent males. Survivor was born into a "family unit" of closely related females, the oldest of whom was the matriarch. She was more than fifty years old and led the group to traditional feeding areas and watering points. A female calf born into a family unit will usually remain in it for the rest of her life unless the group becomes too large. Males, on the other hand, leave the group when they are ten to fifteen years old. Sometimes they wander on their own, sometimes they form groups with other young males.[1]

After eating all the fruits they can find in this grove, Survivor and his group move to a floodplain along the stream, where they feed on the tall elephant grass. Only their gray backs show above the grassy plumes as they pull up bunches of stems and eat the tender shoots. By late afternoon they become thirsty, but do not go to the river because the poachers know the elephants' watering spots and often ambush them when they come down to drink. Since the poachers do not shoot at night, the elephants wait until dark to quench their thirst.

Maybe Survivor can remember the days when his family unit went to the river every afternoon to drink and play. He and the other youngsters would frolic and splash in the water, while the adult females used their trunks to spray their broad backs. But it is no longer safe to linger there. After dark Survivor's group goes to the water's edge and drinks quickly, looking around frequently and listening. They move away immediately and return to the thick vegetation where they cannot be seen.

Day after day, the group feeds on the fruits, small trees, and grass, walking along familiar paths and drinking in the safety of darkness. There is plenty of grass left, and although it is drying along with the season, it is still nutritious.

In mid-June the sky dims from the smoke of the first wildfires started by the poachers. With nothing to stop them except the rivers, the fires sweep across the plains and hills, gobbling up the vegetation that would feed the entire elephant population for months.

Forced by the fires to move on, Survivor and his group turn west and trek toward the scarp mountains. Many of the streams are now dry, so the elephants often walk on the parched riverbeds hemmed in by steep banks. Their large feet leave readily identifiable tracks in the sand.

One afternoon there is a rustling near a tree just above Survivor.

He whirls in alarm, lifting his trunk. One of the other elephants backs into him as they all lurch about in confusion, holding their trunks high to take in the scent. They watch for signs of men with guns. Then they turn and run along the streambed, their feet kicking up sprays of sand. But they are trapped by the banks and cannot escape. Eventually they find a gully and scramble to the top, their sides and rumps bumping one another. Survivor pauses briefly and looks back to see a small troop of baboons climbing into the lower branches of the tree where he had heard the noise. He stops. It's okay. This time.

When the elephants reach the rocky foothills, they often follow the well-worn paths their kind have used for generations. Along the way they feed on scrub mopane and small combretum trees. The trails continue over the mountains, winding around the steepest peaks and into deep ravines. The grass in the mountains is not so plentiful, nor is the water. The group feeds on small trees of the miombo forests, twisting each plant off at its base. They drink at clear springs and streams tucked away in the creases of the range.

On the other side of the small mountain, also walking an ancient trail, is the matriarch One Tusk and her family unit of females. Well over thirty years of age, she leads Misty, Mandy, and Marula — three young adult females — a three-year-old calf, and an infant. Halfway up the mountain they reach the meadow known as Elephant's Playground, where a few palm trees tower over a small stream. The elephants fan out and feed, staying within twenty yards of one another. Misty, who may be the daughter of One Tusk, accidentally backs into the matriarch, but neither of them moves. With their huge backsides lightly touching, they continue to pull up the grass and stuff it into their mouths. Often the females reach out their trunks to sniff each other's faces, or lean against their neighbor. The calf lies flat on the ground, sleeping near the front feet of his mother. Now and then she reaches down with her trunk and moves it along her baby's head.

Soon the three-year-old swaggers over to the small calf and plops down on his rump. The calf lifts his head and wiggles his bum out from under his playmate, who sinks to the ground. The babe staggers to his feet and the two youngsters entwine their trunks, gently pushing against each other. The calf turns away and runs through the grass, his ears and trunk flopping up and down. Within seconds the three-year-old catches up and lays his trunk over the calf's back. They face each other again and push their heads together in miniature sparring.

Abruptly the elephants stop feeding and playing, to listen. A rumbling sound drifts across the clearing from the north, and One Tusk's group returns the call. The temporal glands on the sides of their faces begin to seep, the liquid streaming down their cheeks. One Tusk gives a loud, short rumble and they walk quickly northward. The mother softly prods her infant with her trunk, and he follows the adults in a half-run.

In the trees beyond the meadow, Long Ear and her small family unit emerge at a trot from behind an outcropping of boulders. Their ears held out and rumbling loudly, they rush toward One Tusk and her group. The greeting elephants swarm together in a confusion of purring, twisting of trunks, and clanking of tusks. All of their faces are streaked with the temporal gland secretion. One Tusk and Long Ear wrap their trunks together and flap their ears vigorously; the others do the same, all the while rumbling loudly.

Long Ear, whose left ear has been torn off at the bottom, making her right ear look long, is the "child matriarch" of her group. She is only in her mid-twenties, not old or experienced enough to be a matriarch (in stable elephant populations, matriarchs can be as much as fifty or sixty years old). But the three older females in Long Ear's group were shot by poachers. Now she and the two other young females — one is probably her sister, one her daughter — roam together. They have no young. Last year one of them gave birth, but the small, squiggly baby died the next day, never having found the teat of her inexperienced mother. Sometimes Long Ear's family joins that of One Tusk, and they forage together. The two units make a "bond group"[2] and probably are all closely related.

Groups of female elephants are not haphazard formations that simply bump into one another in the bush. They are close-knit families of relatives whose kin lines are generations old. They communicate with a variety of vocalizations — rumbles, trumpets, screams — except that in North Luangwa they rarely trumpet, apparently afraid that they will betray themselves to poachers. Odors in the secretions of their temporal glands contain important social messages, but they may communicate most by touching. They usually stay within thirty yards of one another and often reach out their trunks to stroke, caress, or sniff their kin mate.

One Tusk gives a loud rumble and the two groups move a short distance into a thicket of miombo woodlands, where they calm down and stand napping in the midday heat. The elephants are quiet and still. Now and then a tail swishes at a fly; now and then a trunk is lifted and sniffs gently along the face of a sister.

In days gone by, the family units and independent males continued over the mountains to the plateau beyond, grazing the lush grass of the extensive glades and dambos. But now these areas are cultivated by man, and poaching is intense. So the elephants must remain in the mountains feeding on the small trees during the months of June, July, and August. Sometimes Survivor and his group come upon small family units of females. Occasionally they approach the females to feed nearby or to check if any of them are in estrus, but most of the time the males are on their own.

In September — the heart of the hot, dry season — Africa performs a miracle. Long before the first raindrop falls, many of the trees, large and small, burst into growth as green and tender as spring. While the grass is still parched by the fires or dried by the sun, the leaves of almost all the trees and scrubs choose this moment to unfold. The valley and mountains are covered with this new life that is so fresh and bright that it seems to glow. And once more the elephants begin to move.

Following many of the same paths down the mountains, Survivor's group and the family units walk back across the foothills toward the river valleys. They pass again through the belt of marula trees, but now there are no fruits. They feed instead on the abundant new leaves and seedpods of the mopane, combretum, and terminalia trees. Since the rains have not begun, many of the rivers and water holes are still dry. The elephants, who must drink every day, are forced to find water wherever they can. One Tusk and Long Ear move along the Mwaleshi floodplains; some elephants forage along the Mulandashi River; Survivor and his group stay in the area of the Lubonga.

In late October and early November, the legendary Luangwa storms build towering monuments of cloud in the sky. Windstorms and sandstorms slap and tease the valley, now trapped in a stifling heat. Some of the most spectacular lightning displays on earth flash across the silent savannas. Many of the new leaves have begun to wither and droop, as though their early burst of energy was too optimistic. All of life, both plant and animal, seems to pause as if waiting. Then in mid-November the first rain falls. Almost immediately the elephants, wherever they are — along the Mwaleshi, near the Lubonga, or still in the foothills — begin moving slowly toward the plains.

Survivor, followed by his four companions, crosses the Fitwa River near Mvumvwe Hill and walks east through the mopane forests. Once in a while they stop to feed, but mostly they keep moving. By the time they reach the plains, miles of young green grasses march across the savannas in endless parades. Survivor sees massive herds of fifteen hundred buffalo and several hundred zebras grazing the new grass, which is surging with more nutrient per volume than almost any plant in the valley. He watches other elephants walking onto the plains from the west, south, and north. Not all of them have migrated to the mountains; some have moved south and north along the Luangwa River. But now most of them — small groups like his own, solitary males, and family units of females — move out onto the expansive plains.

By mid-January about 80 percent of the North Luangwa elephant population has assembled on the grasslands. Even though three-quarters of them have been slaughtered by poachers in the last fifteen years, it is still incredible to see three thousand elephants strung out in herds along thirty miles of plains. One Tusk and Long Ear's females join other bond groups to form large aggregations. Perhaps the grass is too tempting or perhaps they feel safer in numbers, but the elephants leave the sanctuary of the tall grass and feed in the open. Almost like the old days, herds of a hundred elephants stroll gracefully through the grass in long gray lines.

One day, feeding on a soggy plain, Survivor sees a female elephant running at full speed away from several males on the other side of the savanna. Instantly he and his four companions run toward the commotion, their feet sloshing and sucking up the mud. When they arrive, eight young males in their twenties are pursuing the female, who is twisting and turning as nimbly as an elephant can, to escape. One male finally catches the exhausted female and, placing his trunk over her back, attempts to mount her. Another male rams him in the side with the top of his head, and the female dashes off again.

This chaotic scene is not necessarily the way elephants mate. Before poaching was so intense, when a female came into estrus she would make every effort to avoid the young males until the arrival of a musth male — a fully mature, sexually active male more than thirty years old. The two of them would form a consortship for three to four days, during which he would guard her from other males. The pair would mate occasionally and feed together in a relatively peaceful setting.[3]

But most of the musth males in North Luangwa have long ago been shot by poachers. Survivor has not seen one in several years; perhaps all of them are dead. Without a musth male to protect and mate her, the female has no choice but to succumb to these inexperienced bullies. During the next four days she is mated by five different males. She spends most of her time trying to escape them and rarely has a chance to feed. It is not certain that she will conceive under these conditions, and even if she does, it will not necessarily be by the best and strongest male.

Even though aggregations of elephants moving across the plains may resemble the great herds of yesteryear, they are not the same. An elephant's ivory grows during all its life; so does its wisdom. Most of the musth males and matriarchs are dead, and along with them much of the knowledge, experience, and memories of elephant society. This younger generation carries on in the tradition of the past as best it can, but the social system seems in large part to have died away with the numbers.

If the rains are heavy, the plains become waterlogged by February. Survivor's group moves westward to the fringes of the savannas. They make forays onto the plains as the rains allow, still feeding mainly on the lush grass and its nutty-tasting seeds. But almost with the last raindrop, the grasses dry and wither. By April, Survivor is on his way again toward the mountains. There is compensation for the drying grasses; soon the marula fruits will ripen and fall to the ground. And Survivor knows where the marula trees are.

10

Eye of the Dragon

MARK

Late afternoon. Distant shouts.
Young raw voices, male, floating
In the heat. Are they angry, or
Bored, or is it the heat shout
ing through them?
You forget where you are sometimes,
Where you started from.

— JOYCE CAROL OATES

o o o

WE SIT HIGH UP in our Unimog, in the early dry season of 1988, eyeing a pole bridge in front of us that sags across the deep stream cut like a wet spider's web. In the back of the Mog are bicycles, sleeping bags, mosquito nets, camping mattresses, boots, first-aid equipment, food, and other supplies for the men of the Nsansa-mina and Lufishi game scout camps.

This equipment is not coming a moment too soon. The scouts must start patrolling. Having wiped out most of the animals at the fringes of the park, the poachers are striking right at its center. North Luangwa is bleeding from the heart. Each volley of gunfire that we hear, and each cloud of vultures that we see, reminds us that the last of the elephants are dying.

By comparing our aerial wildlife censuses with one flown in 1973 by a team from the United Nations Food and Agricultural Organization, we estimate that poachers have already killed more than twelve thousand of the park's seventeen thousand elephants, about three of every four, and a thousand more are dying each year. Since 1973 between seventy-five thousand and one hundred thousand elephants have been poached in the Luangwa Valley as a whole; *that's roughly one for every word in this book.* Perhaps twenty thousand to thirty thousand elephants are left in Luangwa, and no more than five thousand in the North Park. At this rate they will all have perished in four to five years.

Before we can expect the Mano scouts to go after poachers armed with military weapons, we must equip them properly. We are hoping this Mog-load of supplies will help motivate them.

The Mog weighs more than six tons empty. Its load of tools, winching tackle, camping gear, and game scout supplies brings its total heft to almost eight tons. The poles that make up the floor of the bridge are no thicker than my calf; they span the stream, thirty feet across, bank to bank. A few smaller limbs laid crosswise on top will help distribute the truck's weight. Even so, nearly four tons will come to bear on the poles under the wheels on each side as we drive across. From my seat I can see through the bridge to the stream, eight feet below.

"If the Mog breaks through this pile of poles, we're going to have a tough time getting it out of the stream," I say, leaning my elbows on the steering wheel. "And if one side breaks through and not the other, it'll roll off and end up on its top in the water." Delia and I climb down from the cab and search for several hundred yards upstream and down for another way across, but there doesn't seem to be one.

With a stick from the woods I measure the distance between the Mog's two front wheels, and compare it with the width of the bridge. The truck will barely fit, and the two right wheels will bear fully on a single pole on the upstream side of the bridge. Meanwhile the left wheels will track back and forth between the outside pole and the one next to it on the downstream side.

I scramble down the steep bank and examine the underside of the bridge, looking for rot. The bark on the timbers disintegrated long ago and the wood is peppered with holes from sawdust beetles. Other than that they look fairly sound, though none of them appear strong enough to support the Mog. Nevertheless, we will have to give it a try. If we can't get these supplies to the scouts, there is no hope of protecting the elephants.

Delia signals me into alignment with the narrow bridge, then wades into the stream to keep her eye on the poles as the Mog's weight comes to bear on them. At the first sign that one is giving way, she is to wave me back, although there will probably be little warning before the truck breaks through.

I shift to low-low and creep forward at less than one mile per hour, watching out the window to align my right front wheel with the extreme right edge of the bridge. The heavy wheel finds the butt of the outside pole and drives it into the ground as the truck crawls forward.

Pow! Crackle! Snap! The poles complain as they bow under the weight. I stamp the clutch pedal down, hit the air brakes, and look down at Delia. Peering at the sagging underbelly of the bridge, her jaw rigid, she is waving me forward.

I open the door and jump down to look for myself. The truck's front wheels are already warping the poles badly, but they do not seem to be splitting yet. With the turbo diesel at dead idle, I let out the clutch and the Mog creeps forward.

As I reach the center of the bridge, more shots ring out from the overstressed wood. Ka-pow! The shattered end of a pole flies into the air and the Mog lurches to the left, swaying back and forth, up and down, on the bridge — which seems to be trying to catch its breath. I put my hand on the right door handle, ready to jump clear if the truck rolls left off the bridge. If it rolls right, toward the driver's side, I will have to stay inside.

"Stop! STOP!" Delia shrieks, waving her arms.

The Mog sways as though on a rope bridge. Its left front wheel has broken one pole and is forcing two adjacent ones apart. I sit quietly for a few seconds while the bridge settles down. Then, with Delia signaling, I switch the dashboard toggle to all-wheel drive, turn the steering wheel left, and reverse. Aided by traction from the other three, the left front wheel crawls slowly back up onto a pole. But now the right front tire is half off the right side of the bridge.

I see no sense in sitting at the center of the span waiting for the thing to collapse. I ease the Mog forward, its engine growling amid the gunshot sounds of the breaking bridge. As we teeter along the outside poles, it sounds as if we are driving over a bed of fire-crackers. At last the truck's front wheels reach solid ground and, gunning the engine, I pull clear of the matchstick bridge. After their mauling by the Mog, the poles are even more of a jumble. We will have to rebuild parts of the bridge to get back to Mano. And someday soon we will have to make a proper crossing here.

We camp for the night in the cool, sweet brachystegia woodlands along the stream. Beneath a full moon we swim in the clear water, drifting with the current among moss-covered boulders. There are no crocs to worry about in this montane climate high above the valley — or hippos, for most have been hunted to extinction. We hang a mosquito net from the tailgate of the Mog and sleep with our heads in our backpacks to keep spotted hyenas from biting our faces.

The next morning we "mog" along a track that according to Island Zulu has lain unused for two decades. It isn't a track really, just a path of lesser resistance along which there are smaller trees, an occasional old stump, and slightly taller grasses than in the surrounding forest. Using the Mog and its bull-bar, we bulldoze brachystegias and julbennardias up to six inches thick, crowd past heavy branches, climb over logs, and cross more pole bridges. Although it is tough going, the temperature is pleasantly cool, with fog swirling through the trees and the sun showing through occasionally as a vague silver disk above us.

It is still early when we roll into the Nsansamina camp — little more than three small mud wattle and thatch houses set on a bare earthen clearing in the forest. The fog is lifting; the sun streams rich and golden through puffs of white cloud to the verdant forest below. Three scouts are sitting on squat, hand-carved stools around a small campfire. Its smoke curls up to the blue sky, like a gray rope climbing through the still morning air. One of the scouts is tending a steaming pot of sweet potatoes with a bright green banana leaf for a lid. The other two men are playing musical instruments — one, a thumb piano; the other, a sort of single-string guitar with a gourd base. Behind the men is their fowl's roost, a miniature thatched rondavel set on stilts with a ladder leading to the door. Rubbing sleep from his eyes, the fourth guard stumbles from one of the huts as I switch off the truck.

"Mashebukenye, Mukwai!" We shake hands with Patrick Mubuka, the Camp-in-Charge, and each of the three scouts. For the next few minutes we squat at their fire, inquiring about their health, that of their families, and any problems they might have. Their problem is survival. They have too little food, no medicine, no transportation, no backpacks or camping equipment, and very little money. They have two, maybe three, rounds of ammunition for two rifles that half-work, if they work at all.

"Sa kuno — come over here, please. We have some things for you." The scouts assemble next to the Mog as I climb aboard and begin tossing down boots, sleeping bags, camping mattresses, mosquito nets, first-aid supplies, and even bicycles, which came all the way from the United States. Instantly they are the best-equipped scouts in Zambia. None of the others have even a full camping kit. They clap their hands in thanks.

I am busy assembling a bicycle when a small boy with a big grin sprints past the Mog. Proudly stretched across his chest is a T-shirt with the menacing face of a bulldog and "Go, You Hairy Dogs!" stenciled on it. An instant Georgia fan! On his heels is another child in a shirt with bright red flowers and yellow designer pants decorated with flamingos. Squeals of laughter draw me to the fowl's roost, where Delia is surrounded by women and children. From a big cardboard box she is dispensing clothing, medical supplies, coloring books, and crayons.

The last thing out of the box is "Luangwa Lion," a puppet who tells the children he needs their help to conserve the animals of the valley. The lion explains that he and the other lions live together in communities, similar to villagers in a chiefdom. These communities of lions and other animals are becoming more and more rare, because many men are killing them. So few are left in Africa that people all over the world consider those in North Luangwa priceless natural treasures.

The lion tells them: your fathers have a very important job to do: they must protect all of us animals in the valley from poachers. Remember, you can shoot an animal only one time and then he is dead and gone forever. His meat and skin can be used only once.

But if you keep us alive, you can show us to tourists over and over again and each time they will pay to see us. We are worth more to you alive than dead.

As we are climbing into the Mog to find a campsite for the night, the middle-aged wife of one of the guards calls to me. She cannot speak English, so she takes me by the hand and tugs me toward one of the small huts nearby. At the doorway I duck under the thatch into the dark interior. Even before my eyes adjust to the gloom, I can hear death on the breath of her child. Calling to Delia to bring a flashlight, I kneel down beside the girl. She is perhaps twelve years old. As I reach out to put my hand on her forehead she draws back, her eyes round with fear.

"Mararia," her mother says matter of factly; then to the girl, "Owensee — doctor." And the girl lies back on her grass mat.

Delia arrives with a flashlight and our medical kit. I switch on the light and look into the girl's yellow eyes as she pants with the fever, her lips parched and cracked. I feel for her liver; it is like a lump of cork under her hot skin. The girl will be dead within hours unless we treat her now, although neither of us is a medical doctor.

"Delia, go to Patrick Mubuka. He's the only one who speaks English. Tell him to get the women to bring cool water and some cloths. We've got to bring her fever down right away or she isn't going to make it."

While waiting for the water, I dig out a syringe, a vial of soluble chloroquine hydrochloride, and some aspirin to break the fever. When Delia returns we put wet cloths on the girl's forehead, hold her up so that she can drink and take the aspirin, then I give her an injection of chloroquine.

"If this is chloroquine-resistant malaria, she's probably not going to come right. If it isn't, she has a chance," I say to no one in particular. I pat the girl's arm as we turn to leave, but she does not respond. I should give her some chloroquine pills, but we have run out of them. Before driving away I promise the mother, through Mubuka, that I will bring some more medicine soon.

The next morning we set off toward Old Lufishi — and run into a wall of thick brush only four hundred yards beyond Nsansamina. We've had enough of bush bashing, so we drive all the way back to Mukungule and hire fifteen ax men to clear a fifteen-mile track between the two camps and to build another stream crossing.

While the track is under construction, we drive back to Marula-Puku and return with medicine for the sick girl at Nsansamina. Nine days later she has recovered, the new route is open, and we continue our journey. At Lufishi we give out more sleeping bags, T-shirts, and prophecies from the lion puppet, then return to Marula-Puku to rest and resupply before visiting Mwansa Mabemba and Chilanga Luswa camps.

o o o

We have been away for the best part of three weeks, sponging off in cold streams and the Mwaleshi River as we drive camp to camp delivering equipment to game scouts and their families. Both of us are looking forward to a hot bath as soon as we get home. Along with the gear we hauled to Zambia, we brought a bathtub that I have recently mounted in rocks in the corner washroom of our bedroom cottage. This will be our first opportunity to use it. At the workshop I pull on the Mog's airbrakes, switch off, and while the guys unload our gear, Delia and I grab four kettles of hot water from the fire at the kitchen and head down the footpath to the bedroom.

I set my kettles aside, light two kerosene lamps, hand one to Delia, and open the door so that she can enter the dark cottage and bathe first. She shuffles around the stone wall to the washroom, feeling her way in the lantern's dim light. I have just closed the outside door when I hear the clatter of Delia's kettles on the floor, followed by a screech. Before I can react she sprints out of the washroom, nearly knocking me over.

"Mark! There's a lizard in the bathtub!" she quavers.

"Is that all? I thought you'd been bitten by a snake. What's the matter — you like lizards."

"I do," she says, "but this one is as long as the tub."

I switch on a flashlight from our dressing table and inch quietly around the wall into the washroom, aiming the beam at the bathtub. Two red, beady eyes glow above a huge, blunt reptilian snout. A blue, forked tongue flicks toward me like a bolt of lightning. The thing grasps the edge of the tub with scaly feet and lunges at me, hissing like a ruptured steam pipe, its long dragon's tail lashing about.

I am reversing at high speed when I bump into Delia, who is watching from the doorway. Safely back in the bedroom, I break out laughing. "Poor Boo." I try to be sympathetic. "Most women worry about finding cockroaches in their bathtubs. But you have a five-foot Nile monitor lizard nesting in yours. The tub is full of grass straw! I think it's Mona, the one who was hanging around while we built the cottage."

"I don't care who she is. How did she get in, and how are we going to get her out?"

We finally conclude that the mother-to-be must have crawled through the small window above the tub. I close the door to the washroom, go outside, and lower a tree limb through the window into the bathtub, so that she can crawl out again. But she doesn't, at least not right away. Lizards are cold-blooded animals, which means that their body temperature fluctuates with that of the air. It's mid-July, winter in Africa, and outside it's about 45°F., with a brittle wind blowing. She apparently has no intention of giving up her comfortable bathtub nest.

Armed with a broomstick, I stalk into the small washroom toward her, to see if I can coax her to leave. As soon as she sees me she begins hissing again and smacking, her tongue moving forward and back in her mouth. I slowly extend the broomstick, to see if she is bluffing. Quick as a wink she grabs the end with the teeth of a miniature *Tyrannosaurus rex*, almost biting it in two. Impressed, I back around the corner into the bedroom.

"Unless you've got a better idea, I think I'd rather bathe outside," I say to Delia. Half an hour later, we warm-blooded animals are shivering and swearing in the frigid wind as we splash water over ourselves from a plastic basin; Mona, the cold-blooded monitor lizard, is snug in our tub. The next morning she is gone, apparently to find a meal before returning to her nest. I remove the tree limb from the bathtub and close the window, feeling a little guilty.

A few days later, on a supply run to Mpika, we stop at park headquarters to meet with Warden Salama. But when we step into his office, a short, stout man with remarkably tiny ears and a large gap between his front teeth is sitting behind Mosi's desk. Grinning shyly, he introduces himself as Bornface Mulenga.

"I am the new warden," he says. "I'm afraid Mr. Salama has been transferred." We ask why, and he explains that since it is a departmental matter he cannot offer details, only that the Mfuwe scouts, from near Chipata in the southeast of Zambia, recently came into Mpika on a sting. They caught Mosi acting as a middleman, smuggling tons of tusks from the North Park to Lusaka. Delia and I exchange glances, amazed not only that the game warden has been charged with poaching but that he has not been fined or jailed, merely transferred to another post. With a corrupt warden at the helm, no wonder the scouts have not been patrolling, and have even been poaching themselves. It no longer seems strange that we have seldom seen a wild animal within ten miles of the Mano camp: most have been shot by scouts, or by poachers cooperating with scouts.

Mulenga seems eager, honest, and capable. He goes on to tell us that the department, in response to our support, has decided to upgrade the five scout camps along the park's western border and merge them into a single law-enforcement entity, the Mano Unit. The unit leader will be John Musangu, who according to Mulenga has a reputation for being tough on poachers at Mfuwe, his previous post. More men will be brought in, more houses will be built to accommodate them, and tracks will be opened to service the camps. Given that Zambia is hamstrung by one of the worst economies in the world, these are generous gestures. We promise Mulenga that we will help in any way that we can.

But the new warden follows this with bad news: "I have had to transfer Gaston Phiri," he says. "He was causing trouble at Mano, flirting with the other men's wives." We are surprised and disappointed. Although Phiri seldom if ever patrolled the national park, at least he occasionally took his men on sweeps through the villages to look for illegal firearms. He was one of the few scouts who showed any willingness or initiative in law enforcement.

We tell Mulenga that the scouts refuse to patrol unless they have at least six rifles per squad; so we have collected the few serviceable guns scattered among the five scout camps and concentrated them at Mano. He gives us four more firearms belonging to Mpika scouts, who never patrol anyway. Mano will now have thirteen rifles. By joining forces with men from Nsansamina and Lufishi, they will have enough manpower and rifles to field two patrols, with one armed scout left over to guard the main camp. After loading the rifles into the truck, we drive back to Mano.

As we give the scouts their guns, we announce new rewards for every poacher convicted and for each firearm and round of ammunition taken in the national park. If a patrol captures even five poachers, each scout will earn an extra month's pay. The money offered by the new warden to build houses for the scouts has curiously disappeared. So we hire poachers from Chishala and Mukungule to do the job.

"This is a very fine thing you have done for us, Mr. Owens. Ah, now you shall see us catch poachers!" exclaims Island Zulu as we shake hands with the scouts. Before we begin the three-hour drive back to Marula-Puku, Zulu leads me to his private sugarcane patch, where he cuts a very large stalk for us to chew during our journey. When we finally drive away from Mano, all the scouts and their children follow us to the river crossing. Standing on the far bank, they wave until we are lost among the hills and forests of the scarp. At last, now that we have equipped the scouts, improved their camp, given them guns and incentives, we feel we have a chance of taking the park back from the poachers.

○ ○ ○

A ring of fire like a dragon's eye leaps from the dark woodland below my port wing tip. I pull the plane into a hard left bank and shove its nose into a dive. As we near the blaze I can make out about thirty individual fires. It looks as if a small army is bivouacked at the edge of the woodland. Poachers with a camp that size will have sixty or seventy unarmed bearers and two or three riflemen, each armed with military weapons — Kalashnikov AK-47s, LMG-56s, G-3s, and others. They could easily shoot down the plane.

With me is Banda Chungwe, senior ranger at Mpika. We have been on an aerial reconnaissance of the park to plan roads and firebreaks. I have purposely delayed our return to the airstrip until the last few minutes of daylight, so that we can spot any meat-drying fires the poachers may be lighting along the river. And seeing poachers in action may light a fire under Chungwe. Although he is in charge of overall field operations, this meek and mild-mannered man has never ordered the scouts on a patrol or done anything to inspire or discipline them. According to them, until today he has never even visited their camps or the national park. They need his leadership badly.

I pull off the power and drop the plane's flaps, checking my altimeter and taking a compass bearing on the fires. Then I take Zulu Sierra down over a grassy swale below the canopy of the woodland. Flying ten feet above the ground, I track the trail of gray smoke from the dragon's eye. Just before the poachers' camp, I check back on the stick, nip up over the trees, and chop the power. The camp with its blazing fires is in front of us and a little to the right. I drop the starboard wing and ease on a bit of left rudder, side-slipping the Cessna for a better view of the camp below — and so that any gunfire will, I hope, go wide of its mark.

I can see several dozen men hunkered around the fires, and three large meat racks made of poles — one at least ten feet long and four feet wide. On these giant grills, huge slabs of meat are being dried over beds of glowing coals. Nearby lies the butchered carcass of an elephant. Its dismembered feet, trunk, and tail have been pulled away from its body.

"You sons of bitches!" I swear. The senior ranger says nothing.

Less than two seconds later, the camp is behind us. In the shadow of the escarpment darkness is falling quickly after the brief dusk, and our grass airstrip is not equipped with lights. I turn the knob above my head and a dim red glow illuminates my flight and engine instruments. I am beginning to get the itch on the bottom of my feet that comes whenever I am pushing things a bit too far in the plane. I pour on the power and head up the Mwaleshi to its confluence with the Lubonga, then follow it to the lights of Marula-Puku. If I do not look directly at the airstrip, I can barely make it out on the back of a ridge a mile northwest of camp. We have about two minutes of dusk left, just enough time to go straight in for a landing. I begin to relax a little.

The strip of fading gray grows larger in the plane's windshield. At three hundred feet I switch on the landing light again. At first it blinds me, but finally some faint greenery begins to register in its beam. The crowns of the trees off the end of the runway look like broccoli heads, growing larger and larger.

Two hundred feet, one hundred fifty, one hundred, fifty. As we soar over the end of the strip, I cut the power and bring the plane's nose up to flare for the landing. All at once, big green eyes reflect in the landing light. Just ahead, a herd of zebras is grazing in the middle of the runway. A puku scampers under the plane, inches from its main wheels.

Ramming in the throttle, I haul back on the control wheel. The stall warning bawls as Zulu Sierra staggers into the air, just clearing the backs of the cantering zebras. Biting my lip, I force the plane away from the ground, banking around for another try at a landing. But in the minute it has taken to perform this maneuver, my view of the strip has been lost to the night.

We are on a downwind leg to an invisible airstrip. I hold our heading for another two minutes, which at 80 mph should put us more than two and a half miles from the runway, then I make a descending turn for the final approach.

I check my watch: the same time and speed should get us back to the field. I force my hands to relax on the controls and continue my descent from one thousand feet above the ground.

Flying blind, I feel my way down slowly and carefully, leaning forward, straining to see the ground with my landing light. My feet are jumpy on the rudder pedals. Five hundred feet per minute, down. Down.

The "broccoli" trees flash below. I still cannot see the air field. I ply the rudders back and forth, swinging the plane and its landing light side to side, trying to pick up the strip.

Two parallel lines snake beneath us. Our track from the airstrip to camp! Lowering the nose, I hold the track in my light until the grassy surface of "Lubonga International" resolves out of the blackness. Pulling back on the throttle, I haul up on the flap lever. How sweet the rumble of the ground under my wheels!

After parking the plane, we climb into the Mog and race down the steep slopes to camp. Delia throws together some black coffee and peanut butter sandwiches, and fifteen minutes later Chungwe and I are in the Mog and battering up the track for Mano. I'm going to need every drop of the coffee, for I've been driving and flying since dawn, bringing the senior ranger to our camp for our two-hour survey flight. But I'm eager. With their new equipment and the senior ranger — second in command only to the warden — sitting next to me, the scouts will have no excuse for not coming on this operation. It is one of the best opportunities we've had to send a strong message to the poaching community.

At 10:30 P.M. we roll into Mano and stop before the new unit leader's squat house of burnt brick and metal sheeting. As soon as John Musangu emerges from the darkened interior of his n'saka, I tell him what we have seen and ask him to get his men ready to come with us. He pauses for a moment, drawing hard on a cigarette, then turns and shouts to the scouts in Chibemba. I wait for Chungwe to add something, but he leans against the Mog in silence. The scouts in the n'saka mill about, while others drift in from distant parts of the camp and begin yelling angrily at Musangu. Finally, he tells the senior ranger and me that they refuse to come unless our project pays them extra for each night they are on patrol. I ask the senior ranger to explain to them that patrols are part of their job, not an extra duty, and they are already being paid for them. Instead, he turns and walks around the truck.

Grumbling, the men glare at me, refusing to move. Nelson Mumba declares that it is too late at night to go after poachers and walks back toward his hut. Neither the unit leader nor the senior ranger orders him to return. Even though as a project director and honorary ranger I have full authority over the scouts, I am reluctant to pull rank on Chungwe.

I cannot understand the scouts' behavior. Maybe they are afraid. "Look, gentlemen," I say, "I've been tracking poachers from the air for a long time. By now the two or three riflemen will have split up, each taking maybe fifteen or twenty carriers with him to shoot more elephants. Most of the men you find at the carcasses will be unarmed; only one or two will have guns. I can land you no more than half a mile from them. And, hey, what about the reward? An extra month's pay for every five poachers you catch."

Finally, an hour and a half after we arrived, and with deliberate delay, Island Zulu, Tapa, and some others collect their new packs and climb into the back of the Mog. Shortly after midnight, with the scouts and their leader in the truck, we begin the rough run back to our camp. At 2:30 A.M. we stop at the airstrip. The scouts pour out of the back of the Mog and into a large tent we have set up for them. I ask Musangu to have the men ready to go by four-thirty. He nods, then disappears under the flap of the tent. Driving on to camp, I slip into bed beside Delia to try to catch a couple of hours sleep.

But I can only lie on my back staring into the dark. At dawn I will try to airlift fourteen men in five round trips from camp to Serendipity Strip near the poachers. In the two and a half years since I last landed there, floodwaters will have littered the runway with ridges of silt, deadwood, trees, and rocks.

Delia and I argued about my making this run just a few hours earlier, when she was standing at the old wooden table in the kitchen boma, slicing bread for my sandwiches. When I told her I planned to airlift the scouts to Serendipity Strip, she stabbed her carving knife into the tabletop with a violent thonk.

"Damn it, Mark! If you go head to head against these poachers, it's only a matter of time before they begin shooting at you — at both of us! They could sabotage the plane, or ambush us in camp or anywhere along the track. We can't fight cutthroats like Chiki-linti by ourselves!"

"Look, I am not going to sit by while these bastards blow away every elephant in the valley!" I jabbed a stick into the campfire.

"But landing at dawn on a gravel bar you haven't seen in two years? You'll kill yourself, and the scouts won't go after the poachers anyway. That's just not smart."

"I'm not going to make a habit of this," I said. "But if we don't do anything when elephants are killed, we might as well not be here. And the plane is our only quick-response tool." The next slice of bread Delia cut was about as thick as my neck, and so ended the argument.

I have barely fallen asleep when the alarm goes off. I roll over, choke the clock, and stare at its face. Four in the morning. I stumble into my clothes and out the door. As I pass the office, I grab my red "life bag" full of emergency food and survival gear, then hurry along the dark footpath toward the kitchen. Delia stands there bathed in a halo of yellow-orange firelight, cooking up whole-grain porridge, making sure that if I die this morning, at least I will be well fed. A few quick spoonfuls, a swig of stiff coffee, a hurried kiss — and I am on my way to the airstrip.

In the Mog I climb the steep side of the ridge below the airfield. Out my window to the east, the sky is sleeping in starlight, soon to be awakened by the dawn. And so are the game guards when I reach the airstrip.

"Good morning, gentlemen," I shout into the tent. "Let's go. The poachers are already on the move." They struggle to their feet, rubbing their eyes, stretching and yawning.

I circle Zulu Sierra, flashlight in hand, untying, unchocking, and checking her for flight. Pulling the hinge pins, I remove the door from the right side and all the seats but mine, so I can squeeze in as many scouts as possible.

"I'll take the three smallest scouts, their rifles and kits on the first trip," I shout to Musangu. "Four more should be ready to go by the time I get back, about twenty minutes after we take off." I will keep the plane as light as I can for the first landing, until I have checked out Serendipity Strip. Still rubbing sleep from their eyes, three scouts shuffle to the plane, shirttails out, boots untied and gaping. I turn back to the plane to do my final checks.

I pull the bolts on each of their rifles to check that they are unloaded, then board the three scouts, seating each on his pack and securing him to the floor with his seat belt.

I climb in, crank up, and begin warming up Zulu Sierra's engine. While waiting for the first light of dawn, I give the scouts their last instructions. "Okay guys, we'll fly low along the river, so that we won't be seen. When I have all of your group there, I'll fly over your heads to show you the way to the poachers. Follow the plane. Get going fast, because they will head for the scarp at first light. While you are moving up, I'll try to pin them down by circling over their camp."

As soon as the stars in the eastern sky begin to pale with the dawn, I take off and turn onto a heading of one five zero.

Seven minutes after takeoff we are slicing through puffs of mist hanging above the broad, shallow waters of the Mwaleshi River. The gravel crescent of Serendipity Strip lies just ahead. I pull off power and put down three notches of flap, going in low and slow, the grass heads clipping the main wheels as I look for anything that might trip us up on landing. The plane's controls feel mushy at such a low airspeed and I need plenty of throttle to keep from stalling into a premature touchdown.

Gripping the control yoke hard with my left hand, I slowly bring back the throttle. Zulu Sierra begins to sink. The grassy, uneven ground, littered with sticks, rocks, and buffalo dung, flashes underneath the wheels. The stall warning blares. The end of the short gravel bar looms ahead. I can't find a good spot to touch down, so I ram the throttle to the panel and haul the plane back into the air.

On my third pass over the area I finally see a clear way through the rubble on the ground. This time I bounce my wheels and the plane shudders. But they come free without any telltale grab that would indicate a soft surface. Once in Namibia I tried a similar trick, and when the wheels stuck in a soft pocket of sand the plane flipped on its nose, burying the prop and twisting it to a pretzel.

The next time around I ease the plane onto the grass. As soon as the wheels are down, Zulu Sierra bucks like a mule, her wishbone undercarriage flexing. I stand on her brakes and we slide to a stop with less than a hundred feet to spare before the riverbank. The three scouts laugh nervously and immediately crowd through the open door of the plane. I grab the first by his shoulder. "Don't walk forward or the prop will chop you to pieces." He nods his head and they are out.

Four trips later all the scouts are at Serendipity. Standing under a tree near the river, they wave cheerily as I fly over, heading for the spiral of vultures and the cloud of smoke that mark the poachers' camp little more than half a mile away. I come in low, dodging the big birds, and side-slipping the plane to avoid being shot. I cannot yet see the camp, but the sickly sweet odor of decaying meat — the honey of death — washes through the cockpit. And there it is, the rack — no, six racks, covered with thick slabs of brown meat — the fires, and the men, at least fifteen of them, naked to the waist, covered with gore.

Flashing over the camp, I yank Zulu Sierra around and drop to the grasstops, bearing down on a gaggle of eight to ten poachers who are running away. I lower my left wing, pointing it at them, holding a steep turn just above their heads. They flatten themselves to the ground and stay put. Still circling, I climb out of rifle range and switch on the wing-tip strobe lights, signaling to the scouts that I am over the poachers. Far below, a spiral of vultures lands on the cache of meat, devouring it in a frenzy.

For the next hour I orbit high and low over the poachers, waiting for the scouts to arrive. Finally, I have to break off and return to camp for fuel. After refueling, I write out messages describing the number and location of the poachers and stuff them in empty powdered-milk cans from the pantry. I add pebbles to each can for ballast and attach a long streamer of mutton cloth. Then I take off again and fly over the camps at Lufishi, Nsansamina, and Fulaza, dropping the tinned messages to the scouts there. The poachers will have to pass through or near those camps to get out of the park. It should be easy for the scouts to cut them off.

By the time I land back at Marula-Puku I have been flying for almost six hours. Numb with exhaustion, I wince as I calculate that I have just burned up two hundred dollars worth of fuel. But it will be worth it if the scouts can capture twenty or thirty poachers.

Others will think twice about shooting elephants in North Luangwa, and maybe some will accept the employment and protein alternatives that our project is offering them. For the first time I feel confident that we are about to make a serious dent in the poaching.

We have no radio contact with the scouts and thus have no idea how the operation is going. Four days later Mwamba and I are fixing a broken truck at the workshop. "Scouts," he says, pointing to a long line of men wending their way toward us along the river south of camp. I shake hands with each of the game guards as they arrive. With them are two old men and a twelve-year-old boy dressed in tattered rags, their heads hanging, handcuffs clamped to their wrists.

John Musangu steps forward. "We have captured these three men."

"This is all?" I ask. "These are only bearers. What about the riflemen?" I have already heard over the radio from the warden's office that the scouts from the other camps have managed to catch only a single bearer.

"They escaped," Musangu declares. He goes on to say that there were fifteen men and a rifleman in this particular group. But except for these three, they all got away. The operation has been a bust — except that the airplane and the vultures have denied the poachers their meat and ivory.

Island Zulu, the gabby old scout, spreads his arms and begins soaring about, purring like an airplane as he mimes the airlift; then, hunching over, he stalks through make-believe grass, parts it, aims his rifle, and fires a shot. Turning in circles, he feigns a tackle of one of the three poachers, now sitting on the ground, scowling and rolling their eyes in disgust. Finally, a twinkle in his eye, he predicts, "With ndeke, now poaching finished after one year!"

I drive the arresting officers and their captives to the magistrate's court in Mpika. The boy is not charged. Days later the two men captured in the operation are each fined the equivalent of thirteen dollars and are set free.

11

The Second Ivory Coast

MARK

Yet, though the hope, the thrill, the zest
   are gone,
Something keeps me fighting on!

— BERTON BRALEY

o o o

"THE ONE THEY CALL Chikilinti talked of coming to this camp to kill you and Madam and to destroy the ndeke," Mwamba whispers as we stand under the marula trees.

"While on leave we were in a bar," Simbeye says, picking up the story, "near Mwamfushi Village, not far from Mpika. Since we are from Shiwa N'gandu, the people there did not know us — or that we work for you. We were there for some hours, standing near the counter, when we overheard four men talking." One of them, about forty-five years old and of medium build, was wearing a brown safari suit, his hair straightened, greased, and slicked back. He walked with a swagger as he moved about the bar. This was Chikilinti.

"Who were the other men?" I ask.

"Simu Chimba, Mpundu Katongo, and Bernard Mutondo," Simbeye continues, spearing a leaf with a twig. "This Chikilinti — the people say last year he went poaching for rhinoceros in the Zambesi Valley with his brother and some others. They ambushed game scouts from Zimbabwe. Chikilinti killed two of them, but the scouts caught his brother and dragged him behind a Land Rover through the mountains until he was dead. Chikilinti escaped by swimming across the Zambezi River back to Zambia. He now stays in Mwamfushi."

Shaking their hands, I thank Simbeye and Mwamba. They cannot know how much their loyalty means to me. They turn to go, but Mwamba hesitates, looking back at me. "And Sir, just before Christmas Bernard Mutondo killed a game guard and wounded three others at Nakanduku."

"Thank you; I'll be careful" is all I can think of to say.

o o o

By late 1988 nothing is working, at least not fast enough. In the past month alone, from the air I have found twenty poached elephants; and there were plenty I didn't find. We are losing the battle for North Luangwa. For more than two years we have done everything imaginable for the Mano scouts, and our rewards haven't changed them. They still go into the park only when we find poachers and fly the scouts right in on top of them. Even then they rarely arrest anyone. The idea of a long patrol is about as appealing to them as a bad case of malaria. We can only surmise that it is more advantageous to them to cooperate with the poachers than with us. Our work with the villagers doesn't seem to be having much effect either; many of those we've helped most are still poaching. Loyal, dedicated scouts are our only real weapon against poachers. But the ones at Mano are hopeless. It's time to try some different scouts and some different tactics.

Maybe scouts from Kanona, a game guard post about one hundred fifty miles south of Mpika, would be a better solution. Some are military trained, armed with AK-47s, and because they are so far from Mpika, perhaps they are less corrupt. But in order to use them efficiently, we need to know exactly when and where poachers are coming into the park. Only undercover agents can give us this information.

o o o

Bwalya Muchisa is the son of Kanga Muchisa, one of the most notorious poachers in Africa. Using an AK-47 — thousands of military weapons are floating around Zambia after Zimbabwe's war for independence — Kanga has shot more than a thousand elephants in Luangwa Valley, as well as uncounted numbers of rhinos. A year ago he was captured in South Luangwa Park, where Mfuwe scouts are very good, and is serving eighteen months in jail. Bwalya, determined to better his father's record, at age twenty-six has already killed sixty elephants. Recently, though, the Mfuwe scouts warned him that if he continues poaching they will see him in a cell next to his father's. We have heard that Bwalya is now in Lusaka looking for a legitimate job, and through a friend we have arranged to meet him there.

We drive to Kabalonga market, a row of shops made of concrete blocks and tin roofs. Out front a cobbler sits under a tree working on a four-foot pile of old shoes. As soon as we pull up in our truck, we are hustled by sleek young men like packs of jackals, hawking ivory necklaces and bracelets, malachite frogs and ashtrays, shiny rings, bangles, and beads. A short, well-dressed young man with a pock-marked face and nervous eyes approaches the truck as I step out. He introduces himself as Bwalya Muchisa. With him is Musakanya Mumba, a handsome twenty-two-year-old poacher, fine-featured, soft-spoken, and dressed in a T-shirt and slacks. Musakanya has hunted with Bwalya, using one of his guns, but he is willing to join us and can be trusted if we will give him a job — so he says. I am a little nervous about hiring poachers, and reluctant to take on two at once. But Bwalya says he cannot do the dangerous work of an informant alone. So they hop into the pickup and we drive a short distance to a friend's home, where we can talk without being seen by poachers at the market.

Sitting in an alcove of bushes in the front yard, I ask, "Why would you give up poaching to work for us?"

"Ah but Sir, you know the animals are finishing. There is no future in poaching anymore."

"Okay, but the information you give me will tell me whether I can trust you."

"Sir, you have nothing to fear from us. We are ready to help you in saving the animals," pledges Musakanya. "And I am from Mwamfushi Village, very near Mpika. Many of my friends are poachers. If you can offer employment to them, like me, many will turn in their guns and join us. I am sure of this. Poaching is hard and dangerous work."

I explain that as soon as Bwalya and Musakanya offer poachers jobs with the project in exchange for their guns, the whole community will immediately know that they are working for us. So I tell them to recruit other reliable informants who will remain under cover.

I also say that I will be very surprised if many of their friends turn in their guns for a job because, unlike Bwalya and Musa-kanya, they have not yet been threatened with arrest. But I give them three weeks to convince their cronies to do just that, and offer a thousand kwachas to each one who does. Three weeks from today we will meet at the hut where we stay in Mpika, and from there go to see those poachers who are ready to join us. I shake hands with Bwalya and Musakanya. We have agreed on their base salary, plus a handsome amount for every piece of information that leads me to one of the commercial poachers operating in the park.

"One last thing," I caution before we part. "Be careful. Because they know you're working for me, your friends may play along but then set you up to get hurt by one of the big operators."

"Ah no, this cannot happen," Bwalya says, both of them laughing. "We know these people and our villages too well. We can tell if they are serious."

At Marula-Puku, three weeks after our meeting in Lusaka, we receive a radio message from the operator at National Parks in Mpika: "Bwalya would like to buy that old camera of yours." Using the code we had agreed upon in Lusaka, Bwalya and Musakanya are asking us to meet them in Mpika that night.

Delia and I throw an overnight bag into the plane and fly to Mpika, where we pick up one of our trucks. At about seven-thirty in the evening a knock sounds at the door of our hut. I open it to find two bedraggled men, Bwalya and Musakanya. Quickly bringing them inside, we greet them with pats on the back. Their bleary eyes and sagging shoulders tell us at once that they have been working hard on their mission. We sit around the table as Delia hands them each a cup of strong, sweet tea and I ask what they have learned.

Bwalya tells us four poachers are ready to work for us. "But they won't come here," he says. "They are afraid of being arrested. They are waiting now in the bush near Mwamfushi Village. You must come alone with us to meet them. We must hurry, or they will become afraid and run away."

I agree; but as we are leaving, Delia grabs my arm, saying, "Mark, let's talk about this, please." I ask Bwalya and Musakanya to wait outside for a moment. As soon as the door is closed behind them, Delia hisses, "If you go with them, you're crazy. You don't even have a gun."

"They've done what we asked them to do; now I have to follow through. Anyway, what choice do I have? If these men really do turn in their guns and join us, maybe others will follow. We have to deal with them in good faith. If I'm not back in two hours, go to the police."

Minutes later Bwalya, Musakanya, and I are driving south from Mpika along the Great North Road. We have gone about two miles when Bwalya asks me to slow down. He stares out his window at the tall grass along the ditch bank.

"Stop! You've just passed it." I reverse, swinging the truck's headlights over the berm of the road until I can see a footpath leading into the bush. Dousing my lights, I ease the cruiser off the road, barely able to follow the path by the faint light of the new moon. Grass as tall as the truck's hood swishes along its sides.

We drive for perhaps half a mile in silence. A tree looms out of the darkness along the side of the track as Bwalya says, "Stop. Wait here."

He jerks open the door and the two men run off into the darkness beyond the tree. Two or three minutes pass, and in the dead silence the pulse in my ears makes me uneasy.

I unlatch my door, drop to the ground beside the truck, and crawl through the grass to some bushes twenty yards away. From here I can escape to the main road if this is a hit.

Several minutes later I hear footsteps on my right, headed for the truck. Someone whistles, and in the moonlight I can just make out six men milling around the Cruiser.

"Bwalya, Musakanya! Is that you?" I shout, keeping my head down.

"Eh, Mukwai."

"Okay, switch on the parking lights — the little knob on the lever by the steering wheel — and have the men stand in front with their hands on their heads, so I can see them."

"Ah but Sir, it's okay . . ."

"*Do* it, Bwalya!" After a minute the lights come on, and I can see all of them, including Bwalya and Musakanya, with their hands on their heads. I stand up and walk to the Cruiser.

Musakanya introduces me to the four newcomers, two of whom are sons of Chende Ende, the headman of Mwamfushi. The other two have worked as carriers for Chikilinti and other poachers in the village. They are no more than eighteen years old.

Bwalya asks me if I can take the four teenagers to our camp. "The people of Mwamfushi have beaten very badly some of those who are working with us," he explains. "Now these boys are very much fearing to go back there. Musakanya and myself, we are not afraid; we will stay in the village and continue to pass messages to you." I agree to take the youngsters with us, and tell them to be at our hut by ten o'clock the next morning.

We all sit in the back of the pickup as Bwalya, Musakanya, and the others describe the poaching in North Luangwa. Virtually everyone in Mpika District eats poached or "bush" meat, which is sold illegally in all the marketplaces and by scores of black marketers along the main road. A dealer, usually several dealers, pays a hunter up front for the numbers and kinds of animals they want killed, they agree on a secret rendez-vous near or inside the park, and then each dealer, or sometimes the hunter, hires fifteen to thirty bearers to stockpile mealie-meal.

The hunters, bearers, and often the dealers meet at the rendezvous, occasionally joining other hunting parties there, forming a combined force of up to one hundred forty. A gang this size may include as many as ten to fifteen riflemen, armed with everything from military weapons to muzzle-loaders homemade from Land Rover steering rods. The muzzle-loaders use a gunpowder of fertilizer mixed with diesel fuel, strikers made from the tips of matches, bark fiber as wadding, and balls of steel pounded into a roundish shape as bullets. The powder, balls, and wadding are carried in an animal-skin pouch slung from the shoulder with a strap. The hunters decide who hunts where, then each takes his party to that area, usually in search of elephants and buffalo, large-bodied animals that carry a lot of meat. This information confirms my observations from the air: these splinter groups, consisting of mostly unarmed men, could easily be captured even by poorly armed scouts, if only they would patrol.

As soon as a hunter kills an animal, he leaves it before vultures or smoke from meat-drying fires can attract my airplane. Often he doesn't even go to the animal he has shot, for fear of being caught with it. Some of the bearers stay behind to cut out the tusks, butcher the carcass, dry the meat, and carry it back to their village or to a truck waiting on some remote bush track outside the park. Meanwhile the hunter joins another group of bearers camped nearby, to kill again. When he has filled his contracts, he leaves the park along a route different from that taken by bearers.

In an average three-week poaching expedition, one of the Chende Ende brothers says, each hunter kills from three to fifteen elephants and a larger number of buffalo and smaller animals, such as impalas, puku, and warthogs. An active poacher makes from nine to twelve such trips in a year. In the village of Mwamfushi alone are at least a dozen commercial poachers. It is no longer a mystery why the North Park is losing a thousand elephants each year, why it has already lost more than 70 percent of its elephant population.

"Do they ever hunt rhinos anymore?" I ask.

"The rhinos were poached out years ago," Bwalya reports without emotion. "They were the first to go, because their horns are so valuable. I haven't seen a sign of one since 1982." The others nod agreement. Reports from the early 1970s had estimated the population at seventeen hundred to two thousand. Now there are no more than thirty to fifty in the entire valley, and maybe none.

"Do you think the game guards will ever be willing and able to stop this?" I ask.

"Sir," Bwalya snorts, "they are not serious people in this work. Many of them are friends with the worst poachers. They regularly drink beer with them, sell ammunition to them, tell them where to find elephants and buffalo. The poachers give them meat and money."

Sitting up on the edge of the pickup, Musakanya asks, "You know Patrick Mubuka, the Camp-in-Charge at Nsansamina? He killed two elephants just north of Marula-Puku last year. Another scout reported him to Warden Salama, but he is still Camp-in-Charge."

"Ah, but you know, Sir, it is not just the game guards. It goes much higher than them," Bwalya says, shaking his head.

I lean forward. "How high?"

"When you arrived in 1986," he continues, "two truckloads of dried meat were taken from the park to Lusaka every week, to the National Parks headquarters and to other officials."

"And ivories!" he exclaims. "They call Mpika the 'Second Ivory Coast.' I know about one truck loaded with 547 ivories, all taken from the North Park in some few months."

"What about the magistrate in Mpika? Is he straight?"

Bwalya grinned and shook his head. "As straight as a bull's prick, Sir. I have a friend, Patrick Chende Ende, a cousin to the headman of Mwamfushi. He was charged with poaching. The night before Patrick went to court, the magistrate drank beer with the Chende Ende family. They gave him one thousand kwachas [thirteen dollars]. Next day he fined Patrick two thousand kwachas instead of putting him in prison for a year, as the magistrate in Chipata does in cases like these." He goes on to tell me that Bernard Mutondo, the poacher who killed and wounded scouts just south of the park, was never even arrested. When later charged with another poaching offense, he was sentenced to only four months in prison. Mutondo used a gun belonging to a close friend of the magistrate; it was never confiscated, as is prescribed by law. In the midst of the elephant slaughter, instead of imprisoning poachers, the magistrate is fining them an average of thirteen dollars! And he confiscates only the most dilapidated muzzle-loaders, returning the better guns to the poachers to be used again and again. Twice Mfuwe scouts have even caught poachers using the magistrate's own gun.

"There is more," Musakanya offers. "A very strange thing. Often lately we have seen the new warden, Bornface Mulenga, drinking in Mpondo's Roadside Bar on the main road near the airstrip, that very one where Chikilinti, Simu Chimba, and the others are usually found."

"Okay, guys," I caution them, "don't let your imaginations run away with you. Mr. Mulenga is probably just conducting some of his own investigations. I think he's straight. He has to be straight."

"Sir, you cannot know the extent to this problem," Bwalya adds. "The police in Mpika, Isoka, Chinsali, and the officers at the armory in Tazara, they all give weapons to the poachers. And the army — I know a soldier who brings AK-47s and ammunition from the barracks in Kabwe for poaching in the valley. He comes to Mpika once each month in an IFA [military] truck and takes meat and tusks back with him. And much more ammunitions are coming from the munitions factory near Kanona. You can't believe it."

But I can believe it. Officers from Zambia's Anticorruption Commission have told me they estimate that one hundred fifty to two hundred military weapons are being used for poaching in the Mpika area, many of them from official armories. Yet since we have been in North Luangwa, to our knowledge not one of these weapons has ever been confiscated from a poacher by local authorities.

Then Musakanya tells me that Mpundu Katongo left four days ago to hunt on the Mulandashi River. And even now, Bernard Mutondo is somewhere along the Mwaleshi shooting elephants and buffalo. "If you fly the Shatangala route through the mountains into the valley, you shall see their carriers bringing meat and ivories back to Mwamfushi," he says. In the beam of my flashlight, Musakanya and Bwalya sketch a map showing six major footpaths leading from villages west of the scarp, through the mountains of the Muchinga, and into the park.

"How are we going to stop this?" I groan in despair.

"Sir," Bwalya says, "it is too big. You will not stop it."

"We will. With your help we can and we will. The world is going to change for these poachers."

12

A Zebra with No Stripes

DELIA

But for such errant thoughts on an equally errant zebra
. . . I might have seen a little sooner.

— BERYL MARKHAM

o o o

SIMBEYE, MWAMBA, KASOKOLA — the Bembas who have
worked for us through the years — smile every morning no matter
what the night has left in its wake. If we need a sentinel, they
shoulder a rifle; if we need a carpenter, they lift a saw; if we need
bread, they build a fire. They are as steady and solid as the Muchinga Escarpment that guards the valley.

Together we dug up three thousand stumps for the airstrip, side
by side we collected thousands of rocks for the cottages, step by
step we hiked to untracked valleys. But now that the camp is long
ago complete, I cannot step out of the cottage without one of them
rushing forward to take from my arms anything I carry. I tell them
that I am getting fat with all this assistance, let alone from the fresh
bread they bake in the black pot.

Not once have they refused a request, even if it meant climbing
to the far reaches of a marula tree to cut down a dangerous limb,
or hiking for days to deliver food to the ungrateful game guards.
Unarmed, Simbeye has run after an armed poacher and tackled
him to the ground. Unasked, Mwamba has brought me flowers.
And Kasokola has carried my heavy backpack for the last mile of
many hikes.

They know that we love to hear about the wildlife, so whenever
we return to camp from a trip, they describe in detail the wonders
we have missed. Very early on, we realized that the longer we were
away, the more fantastic the stories would be. We would like to

add some of their observations to our records, but we can't quite be sure of their accuracy.

Mwamba once told us in great detail about a leopard's killing a puku on the beach, and a crocodile's flying from the water to take the puku for himself. This episode could well have happened, and I started to write it in our "Leopard Observations" file. But at the same time Evans Mukuka, our educational officer, wrote in a long report that he had watched rhinos grazing near the airstrip. When I questioned him about it, he said of course there are no rhinos; everyone knows they have all been shot by poachers. But knowing that we would like to read about rhinos near the airstrip, he added them to his report. And so I did not record the leopard and crocodile incident. Still, we enjoy the stories and look forward to hearing them when we return to camp.

Kasokola, deciding that I need a cook, brings his elder brother, Mumanga, back to camp with him after his leave. Mumanga, a slim, jolly man of forty, brightens our kitchen boma daily with a freshly baked pie or cake. With so much more baking going on, Mumanga decides that he needs an assistant, so I hire Davies Chanda from Mukungule to help with the cooking and other camp chores. Chanda, twenty years old, would rather be in the army than in a kitchen and he marches around camp in stiff-legged military fashion, saluting me whenever I pass by.

This morning very early, Chanda and I leave Marula-Puku for the long drive up the scarp to Mukungule for the grand opening of the Wildlife Shop. Although Mukungule is a fairly large village, it has had no store whatsoever. Anyone who wanted to buy a bar of soap, a bit of salt, or a matchstick had to walk two days to Mpika. When the Wildlife Club that we sponsor came up with the idea of opening a shop, we agreed to purchase the first stock of goods, to help with transport, and to lend them enough capital to get started. Months later, after endless delays in obtaining their trading license, the shop — a small, neat mud hut with clipped thatch roof — is ready for its official opening.

The shop is one of several improvements we have helped bring to Mukungule. The North Luangwa Grinding Mill now grinds maize for the villagers for a small fee, and a weekly farmers' market offers a place for people to sell their produce. The Women's Club sews children's clothes that will be sold in the Wildlife Shop. We visit the school every month to teach both children and adults the value of wildlife.

Chanda and I reach the foothills of the scarp just as the sun greets the golden grasses, which stretch for miles over rocky knolls. Before us looms the solid shoulder of the Muchinga Escarpment, challenging our departure from the valley. A small herd of zebras emerge from the mopane scrub and canter slowly along the track. As I stop the truck to watch, they pause to look back at us. Bold black-and-white stripes against a golden backdrop — they merge into the spindly trees and disappear.

Rambling over small boulders and through dry streambeds, we drive on. A few miles later Chanda says, "Madam, you saw the zebra with no stripes?"

"What, Chanda? What are you talking about?"

"In the herd we saw, Madam, there was a zebra with no stripes."

My eyebrows lift. I saw the zebras and they all looked normal to me. I smile. "Well now, tell me, Chanda, if a zebra has no stripes, is he black, or is he white?"

"I do not know about every zebra with no stripes, but this one, she is a she and she is black."

"Oh, I see. Perhaps she was standing in the shadows and she looked black."

"No, Madam, I have even seen her before. She has no stripes."

"Fine, Chanda, that's interesting. Maybe someday I will see her. Now, please, can you help me decide what to say in my speech to the people of Mukungule at the opening. Even after all of our help, they're still poaching." Together we write my speech as we bounce over the scarp.

When we tire of talking and speech writing, my mind picks up where my mouth left off, and I think. This imaginary zebra with no stripes has made me think of the elephants in the valley who have no tusks. We have seen more and more of them. Tuskless elephants occur throughout Africa. But Mark and I wonder if there is not a greater percentage of them in this population due to the heavy poaching pressure on those with tusks. Since the tuskless ones are less likely to be shot, they have a better chance of surviving to reproduce and to pass the genes for tusklessness to their offspring. We must make an aerial count of them.

I ask Chanda what he thinks of the tuskless elephants. "Madam, as you yourself have seen, we tribesmen of the Bemba and Bisa always carry axes. As a small boy, we carry a small ax; as a big man, we carry a big ax. You can always know the size of a Bemba by the size of his ax. A man must never go into the forest without his ax, for if the way becomes too thick, he cannot pass or he must depend on his friends to cut the way. A Bemba without his ax is not a big man. That must be what the other elephants think of an elephant without tusks."

o o o

The Mukungule Wildlife Shop is draped with a wide yellow ribbon, and buckets of wildflowers sit by the door. Dozens of villagers — adults and children — have gathered in the freshly swept yard. The district governor, who has come from Mpika for the ceremony, gives a speech honoring the game guards; Chief Mukungule with his ancient eyes asks his subjects to stop poaching; and a young student of thirteen delivers the most moving speech of the day, declaring that wildlife is the best chance the villagers have for a bright future. The ribbon is cut, and people buy soap and matches. Life is just a bit easier in Mukungule because of the Wildlife Club. Are we one step closer to winning over the people, or will they continue to poach? Chanda and I, exhausted by all the merriment, drive back down the scarp.

The miombo forests of the Muchinga Escarpment are so thick that we can rarely see the valley below. But at one point the trees stand apart, offering a spectacular view of the sprawling Luangwa Valley all the way to the Machinje Hills in the distance. When we reach this spot, I exclaim — as I always do — at its beauty. "Yes," says Chanda, "at this place you can see as far as you can. When your eyes touch the other side, they are no longer in Zambia."

By the time we reach the foothills, the sun is as low in the west as it had been in the east on our ascent. I can almost imagine that instead of the sun's having moved, someone has turned the valley around in the opposite direction. The grass is just as golden, the sky just as soft; we have missed the hard, bleached colors of midday. "Now, Chanda, if you see this zebra with no stripes you must tell me." I know that the chances of seeing the same zebra herd eight hours later — stripes or not — are extremely remote. But he misses the teasing in my voice and cranes his neck this way and that, searching for the black zebra.

Seconds later, "There she is, Madam."

"Where?" I jam on the brakes, the truck skidding to the side, and look where Chanda is pointing. In full view at the edge of the mopane trees, only twenty yards from us, is a large female zebra. Her face, neck, and body are the color of dark charcoal with only a faint shadow of stripes. Beside her is a small foal with a perfect black-and-white pattern.

Chanda is grinning from ear to ear. "You did not believe me, did you, Madam? But here is a zebra with no stripes."

"You are right! Chanda, you are right! She has no stripes. At least, she has fewer stripes."

The female turns and faces us head on. With this stance her stripes fade away completely. "Madam, is a zebra still a zebra, if she has no stripes?"

"Yes, Chanda, she and all zebras have the genes for many combinations, but in nature usually the stripes prevail because they have advantages in the wild. Like camouflage, for example."

"Well, Madam, that may be so. But for myself, I just don't know what the world is doing, to make elephants with no tusks and zebras with no stripes."

13
Chikilinti Juju

MARK

In my rudyard-kipling-simple years I read
Of mid-jungle where the elephants go to die.
Old bulls know, and rather than death by herd,
Wait alone, and add to the fabulous ivory.
— JOHN HOLMES, "The Thrifty Elephant"

o o o

KASOKOLA AND I MIX a sludge of diesel fuel and sand in a pail, then spoon it into Nespray powdered-milk tins. Each is equipped with a twist of diesel-soaked rag, which will serve as a wick for the homemade flare pot.

At the airstrip, after dark, we remove their lids and set out our lighted flares at hundred-yard intervals along both sides of the runway. I park the Land Cruiser so that its headlights will play across the end of the strip. Before going to the plane I plug a spotlight into the cigarette lighter socket. The light will serve as a beacon to help us find the airstrip in the dark. Musakanya has radioed us that Chikilinti, armed with four brand-new AKs, is somewhere in the park with an army of other hunters and bearers. For several nights we have been flying to look for his fires.

It is early 1989, and for a while now I have been keeping two guards posted on Foxtrot Zulu Sierra. Poachers would relish a chance to strafe or burn it, or at least chop it to bits with their pangas, or machetes. To further guard against this, I have evoked several of my own brands of juju. Some time ago I mounted solar lights on poles set in opposite corners of the plane's boma, each with an infrared beam fixed on the aircraft. Whenever anyone approaches, breaking one of the beams, the lights switch on automatically and suddenly the plane is brightly lit, almost jumping out of the darkness — as if by magic. One day I convinced the game scouts that Zulu Sierra cannot be hit by bullets — by drawing my pistol and shooting at it with blanks. No holes. It must be juju. Another day I took one of the scouts flying and homed on one of our radio-collared lionesses across miles of wilderness; I even talked to her over the plane's radio — or so it seemed. The scout didn't know about homing on radio transmitters and I didn't tell him. The word about my juju has spread quickly. It may work for a while.

Now, as I untie and check the plane, I tell the guards to make sure the flares stay lit. Pulling its hinge pins, I remove the door from the right-hand side of the Cessna so that Kasokola can have a better view of the landscape; then we taxi to the end of the strip and begin our takeoff run. There is something unearthly, surreal, and primal about accelerating to take flight at night with the door off: the rushing wind, the roar of the engine, the vibration, and the lines of flickering yellow flares speeding past.

I pull back the yoke and point Zulu Sierra's nose at the waning, lopsided moon as we take to a starry sky. The wheels lose contact with the earth and their rumble and vibration cease. We seem to glide through a liquid combination of nightness and moon day; through another time-space dimension where neither night nor day prevails; where their elements are equally mixed like two pigments on an artist's pallet, blended to yield an altogether more beautiful hue. This hue, combined with the cool, humid night air, makes for a flying medium that is languid, moist, and dense, like the water in a blue-black pool. We are not flying so much as sailing through this celestial pool, and I can almost imagine that the stars are white waterlilies, or points of phosphor drifting by.

Fifteen minutes later our dream-like flight has led us down the Lubonga to the Mwaleshi, then up the Lufwashi. As far as I can see, Africa is asleep. Even the hills and mountains of the scarp are recumbent in the darkness. I bank the plane across the shadowy face of Chinchendu Hill, then back toward the Mwaleshi River. Like a silvery snake, the moonlit waters gradually resolve out of the night. Minutes later, Kasokola's faint voice finds its way through the roar of the wind and the engine into my earphones.

"Fires! There!" He points. At a thousand feet above the ground we are nearing a gap in the ridge through which the Loukokwa River flows on its way to the Luangwa. As we pass over it, I look down into the rough amphitheater that the river has carved out of the ridge. Like a giant hearth, the fifty-foot walls glow orange from dozens of campfires. Another dragon's eye.

I pull back the throttle and take two notches of flap for a descending turn that will head us back past the encampment at a lower height. This time I clearly see several tents set up among the trees along the river. Only Chikilinti and his friends are affluent enough, and bold enough, to shelter in expensive, conspicuous tents.

"Can you see meat racks?" I shout.

"No."

"Then we've caught them in time. We'll have to get the game guards down here early." Tomorrow the hunters will split up, set up several camps in the area, and start shooting elephants. I bank the plane toward camp and a short night of fitful sleep.

Up before dawn the next morning, I send a truck for the paramilitary scouts at Kanona, as planned. On their way through Mpika they will pick up Bwalya and Musakanya, who will guide the patrol and tell me how well the scouts conduct themselves. This crack unit will be armed with their AKs, so going after one of these bands of poachers should be no big deal; they'll give a fitting reception to Chikilinti's group.

Then, knowing it is probably useless but determined to do everything I can, I fly to all the camps around the park, dropping powdered-milk tins with notes telling the scouts to intercept any poachers fleeing from our operation on the Loukokwa River. All of this flitting about the countryside dropping tin cans is still necessary only because National Parks has not, in more than eighteen months, arranged licenses for the radios we bought for the camps. And the same is true for the sixty-one guns that were ready to be shipped from the States two years ago.

Late the next afternoon, squatting on its axles, our truck returns from Kanona loaded with scouts. The next morning I airlift them to an emergency airstrip an hour's walk from the poachers' camp. When they are assembled and ready to go, I tell them how happy we are to have their help, and that it should be easy to find at least one of the several poaching bands that have splintered off from the large group I spotted two nights ago. Shaking their hands, I give the group a small radio and send them off. Bwalya and Musakanya, eager as two young hunting dogs, are in the lead.

As the scouts begin moving toward the poachers' camp, I take off, flying in the same direction to show them the way. They need only follow the plane. I am hoping that we will be in time to stop the poachers before they kill any elephants. But I have flown only three miles when I run into three thick spirals of vultures, and looking down I can see the carcass of a freshly killed elephant at the base of each. Tucked up under the trees near the carcasses are clusters of big racks covered with slabs of smoking meat. The cabin of the plane is immediately saturated with the familiar sweet smell that hangs cloyingly in the back of my throat.

Dodging vultures, I take the plane down and discover two more slaughtered elephants lying within a few yards of each other. Three tents and four flysheets are set up under a tree surrounded by a thicket, and scattered around the camp are jerry cans, pots, pans, axes, ropes, and other gear. Ten to fifteen men are hiding in the thicket with a mountain of red meat at least five feet high and six feet in diameter.

"Chikilinti, Simu Chimba, Bernard Mutondo, and Mpundu Katongo." I grind the names of the worst poachers in the district between my teeth. This time the scouts, real scouts, are right here!

Some of the poachers begin running when they see the plane. I circle over them, very low, pinning them to the ground. After a few passes I break off and fly back to where we have just left the scouts, not a minute and a half away. Over and over I call on the radio, but they do not answer me. I scribble a note on a pad: "Found poachers with five freshly killed elephants approximately three and a half miles southwest of you. Will pin them down with airplane and continue circling with wing down to show you their position until you arrive. Come as fast as you can." I tear off the sheet, tie a white mutton-cloth streamer to it, and drop it to the scouts. When they pick it up and wave to me, I head back to the poachers' camp.

Once again I come in low and slow, trying to see details of the camping equipment or anything that might lead us to these poachers if they escape back to Mpika. The encampment is concealed well, in a small forest of stunted mopane trees. Except for the smoke and the vultures, I might never have found it.

I am passing over the last elephant when a man in a red shirt steps out from behind a tree and shoulders a stubby AK-47. I kick hard left rudder, and the plane skids sideways through the air. Looking out my window, I see his shoulder jerk from the recoil and a puff of gray smoke bursts from the muzzle. Suddenly I am very aware that he is shooting bullets with steel jackets — at my flying biscuit tin with its thin aluminum skin.

I am over him and gone before he can fire more than a couple of rounds. For a minute or more I circle away from the poachers, dizzy with the blood pounding in my ears. The scouts should be there in about an hour, but if the poachers start pulling out now, the scouts will have a hard time tracking them in the thick brush. I have to keep them pinned down. My hands damp on the control wheel, I bank for the camp again, skimming the treetops so Red Shirt will have no more than a second to see me and get off a shot. I keep my speed low, so that if I hit one of the vultures turning above the carcasses there may be enough left of the plane to try a forced landing. I once saw the remains of a twin-engine Aerocommander that had run into a vulture. The big bird had blasted through the windshield like a cannonball of feathers, killing the pilot and copilot instantly, then continued through the cabin, destroying its interior. The remains of the vulture, a blob of red meat the size of a baseball, were later found in the tail cone of what had been an aircraft.

I kick the rudder pedals left and right, zigzagging as I head into the vultures and smoke. A bird is coming straight at me, craning its head and flapping furiously to get out of the way. I pull off power and jink to the right. As the roar of the engine subsides, I hear the pop-pop-pop of gunshots through my open side window.

I circle back and this time fall in with the glide pattern of the wheeling vultures, spiraling with them, holding a steep left turn thirty feet above the main camp. But there is Red Shirt behind a tree, pointing his rifle at me again. To put some distance between us, I spiral up high with the vultures. At fifteen hundred feet, with my wing-tip strobe lights flashing, I circle, waiting for the game guards. It has been more than an hour since I dropped my note to them. They surely can see the plane, and they must hear the shooting.

Another hour passes, two, two and a half. Still no sign of the scouts. I fly back in their direction but I can't spot them. Dodging vultures, aching with fatigue, I continue circling over the poachers until my fuel gauges nudge empty. And then I head back to Marula-Puku, certain that the scouts will arrive soon.

They never do, of course. They never come. Days later they show up at Marula-Puku, asking to be taken back to Kanona. They found only one dead elephant, they say, did not see the airplane, heard no shots, and did not find poachers or any signs of them.

The next morning, after the scouts have gone back to Kanona, Bwalya and Musakanya tell me a different story. The scouts joined the poachers, whom they know well, and spent an enjoyable evening eating elephant meat around the fire.

I write yet another report to the National Parks and Wildlife Services, but do not receive a reply, and again nothing is done.

14
The Eagle

MARK

I am the eagle
I live in high places
In rocky cathedrals
Way up in the sky.

I am the hawk
There's blood on my feathers
But time is still turning
They soon will be dry.

— JOHN DENVER

o o o

MWAMFUSHI VILLAGE: eleven-thirty at night, a week after the busted operation with the Kanona scouts. Musakanya and his wife have been asleep on the floor of their mud and thatch home since about nine. Gunshots shatter the darkness outside. Chips of mud and splinters of wood rain down on them. Throwing his body across his wife to protect her, Musakanya holds her down as she screams. More shots. Bullets punch through the walls, kicking out puffs of dust and clods of dry dirt. Musakanya crawls to the door in time to hear angry shouts and receding footsteps.

Neither Musakanya nor his wife is killed, fortunately, but sooner or later someone will be. The poachers have also shot up the home of the Mwamfushi headman, who is encouraging his villagers to stop poaching; and they have poisoned, not fatally, Jealous Mvula and two of our other men. Clearly, they are upping the ante. If they can do it, so can I. But I will have to do it alone. I've had it with scouts.

At dawn I lift off from the camp strip and fly along the Mulandashi River, following one of the poachers' footpaths that Musakanya and Bwalya sketched for me. The poachers have hit this area hard, because it is miles from our camp. In this poachers' shooting gallery, nearly all the elephants and buffalo have been wiped out in the previous four or five years.

Several times lately, when I have discovered poachers from the air, I have chased them with the airplane until they ran away, abandoning their meat and ivory to vultures and hyenas. It is expensive for a hunter to mount a three-week expedition into the park, and if operating in North Luangwa becomes unprofitable, the poachers will give it up. If I can't get them locked up, maybe I can at least bankrupt them.

Staying low so that any poachers in the valley ahead will not hear the plane, I follow the broad footpath. Banking left and right — watching for vultures, smoke, and meat racks — I fly along the trail through the hills to the large baobab at the confluence of the Kabale and Mulandashi.

Just as I begin to turn away from the tree, something on a broad sandbar along the Mulandashi catches my eye. Leveling the plane, I see two men run across the bar, each carrying an elephant's tusks on his shoulder. Other men sprint beside them, lugging baskets of meat.

I point the nose of the Cessna at the two men with the tusks as they dash across the dry riverbed, heading for thick bushes along the shore. They drop the tusks, turn, and race for a stand of tall grass. I put the plane down "on the deck," my wheels skimming just above the sand.

The poachers glance over their shoulders at the plane bearing down on them, their legs pumping hard, sand skipping from their feet. When my left main wheel is just behind one of them, he looks right at me, and I can see him wondering desperately, and much too slowly, which way to dive. At the last instant I nick back on the controls and the wheel misses his head by inches. I look back to see him plunge headlong into a devil's thornbush at the river's edge.

The other men have wriggled into the tall grass patch on the sandbar. I think it needs mowing. On this pass my Hartzell propeller from Piqua, Ohio, chops the tall stems into confetti. The prop blasts it over the poachers, who are lying face down, hands over their heads. As I pull up and away, I switch the plane's ignition quickly off and on. A tremendous backfire explodes from the exhaust pipe. I haul Zulu Sierra around for another pass, and another, burping her again and again.

For half an hour I make repeated passes over the poachers, reinforcing their terror, extracting from them a price for killing in the park. Then I make a show of landing, rolling the plane's wheels along the sandbar, to make them think I am going to stop and try to arrest them. But if I drop too much speed in this loose sand, I will indeed land — without a hope of taking off again. I will be stuck on this sandbar, unarmed in the midst of a gang of poachers who hate me now even more than they did before. So I abandon my pretense at law enforcement and head back to camp.

Even without any arrests, my air patrols begin to have an impact on the poachers. For several weeks I fly almost every day, diving on poachers along the Mulandashi, Luangwa, Mwaleshi, and other rivers. Musakanya tells me that many of the men from Mwamfushi are refusing to work as carriers, so the hunters are forced to hire from villages farther from their base around Mpika. More often now they have to wait until late afternoon to shoot, drying their meat at night to avoid being discovered from the air. And whenever they hear the sound of the airplane, they quickly cover their fires or put them out with buckets of river water. Other poachers are setting up their meat racks in the cover of hills where I haven't been flying. These air patrols are the only thing protecting the park, so often I fly until midnight and then get up at three to fly again.

Delia is more supportive of these antipoaching flights than of my "suicide runs" to pin down poachers for scouts who never show up. "Finally, something is working," she says, "even if just a little." Still, the park is losing elephants at the rate of five hundred a year — too many, too fast. We don't have long to save the rest. And Zulu Sierra and I aren't going to be able to work our juju on the poachers for much longer. Soon they will realize that the plane can't really harm them. We need help. And at last we are going to get it.

In October 1989, in a warm room in a snowy land far from the shimmering savannas of Africa, seventy-six nations of CITES (the United Nations Convention for International Trade in Endangered Species) vote to list the African elephant as an endangered species (see Appendix B). In so doing they forbid trade in ivory and all other elephant parts, and provide the last hope for earth's largest land mammal.

15
Moon Shadow

DELIA

The only paradise we ever need — if only we had the eyes
to see.
— EDWARD ABBEY, speaking of the earth

o o o

I AM ALONE IN CAMP. Mark is flying an antipoaching patrol and
all the Bembas are working on the track between camp and the
airstrip. In our stone and thatch office cottage, I analyze more
survey data — calculating how many buffalo, impalas, kudu still
remain in North Luangwa. I can hear the Lillian lovebirds whis-
tling as they fly from tree to tree, feasting on the ripe marula fruits.

Weary of numbers, I walk to the river's edge and sit cross-
legged with Nature for a while. Across the water, far away at the
next bend, I see a tall gray bush moving. Then an elephant, also
tall and gray, steps from behind it. It is the first elephant I have
seen from Marula-Puku. In exactly four seconds he disappears in
the tall grass. We have worked three and a half years for those four
seconds. It has been worth it.

When Mark returns, we walk to riverbank and I point out where
the elephant had been. To our astonishment two others appear at
that moment, in the same spot, and wander along the river. All at
once it seems that there are elephants everywhere. The next morning Mwamba runs to the office cottage to show us five elephants
feeding on the high bank behind the workshop. They are the
surviving members of Camp Group — and at last they have accepted us enough to feed within sight of camp. When the Bembas
walk to the airstrip, they see three other elephants feeding on
the tall grass of the distant hills. They have finally come home to
the Lubonga.

A few days later I am alone in the office cottage again, still working on the game census figures. At first I am only faintly aware of the sound. Then I lift my head and listen. It is an odd noise — whaap, whaap, whaap — like someone beating a blanket. I try to work again. Whaap, whaap! I look out of the window but see nothing unusual. Too curious to work, I step out of the cottage and look around. The sound is coming from the other side of the river. I look toward the tall grass some forty yards from camp, and I gasp.

There, standing in full view, is Survivor, the elephant with the hole in his left ear. Apparently relaxed and unafraid, he wraps his trunk around the base of a thick clump of grass, pulls it up by the roots, swings it high over his back, and pounds it on the ground. Whaap, whaap, whaap! Dirt falls from the roots; he sticks the cleaned grass into his mouth and chomps.

I do not budge, sure that if he sees me he will run away. I have been watching him feed for twenty minutes when I hear the truck coming. If Mark races into camp the way he usually does, it will frighten Survivor. I ease along the wall of the cottage and creep toward the road, watching the elephant carefully. Surely he can see me, but he shows no sign of alarm. Once I am around the corner and out of his sight, I race through the grass toward the approaching truck.

When Mark sees me running toward him, waving my arms, he stops the truck and leaps out, shouting, "What's wrong?"

"You're not going to believe it! Follow me." We tiptoe behind the cottage, around the corner, and into the open doorway. Standing just inside the cottage, we watch Survivor feed. He is not a large elephant, but at this moment he seems enormous — he represents hope, success, and a glimpse of Luangwa as it should be.

Eventually, confident that the elephant does not feel threatened, we walk slowly to the riverbank and stand near a marula tree. After some time, Survivor lifts his trunk and walks directly toward us on the far bank. He stops at the river's edge, only twenty yards away, and raises his trunk, taking in our scent. It is an elephant handshake — a welcome, across an invisible line, into the natural world. I feel the honor of the moment deep in my heart.

o o o

We long to follow lions across moonlit savannas, to rise at dawn to see where zebras feed, to count infant elephants standing beneath their mothers' bellies. But the antipoaching battle has devoured our time, energy, resources, even our spirit, since we arrived in North Luangwa. We know that even important scientific observations are a luxury in this place at this time. How can we observe the social behavior of lions when elephants are literally dying at our feet?

Yet at this moment there is a respite; elephant poaching has declined. It is now April 1990, and in the last seven months Mark has seen only five dead elephants from the air. We are convinced that his antipoaching flights and the CITES ban on ivory are working. The price of ivory has plummeted from one hundred fifty to five dollars per pound, and 90 percent of the world's legal ivory markets (plus some illegal routes) have dried up.

Six African nations, however — Botswana, Zimbabwe, Mozambique, South Africa, Malawi, and Zambia — have filed reservations against the ban, and this allows them to export ivory legally or to sell it within their borders. It is appalling to us that Zambia, where in only fifteen years poachers have killed 115,000 of 160,000 elephants, has refused to join the ban. Since there is no legal elephant hunting or culling in Zambia, all of the ivory in this country is from poached elephants. The southern African nations have formed their own cartel for trading ivory, and we are afraid that soon new local markets within this region will replace the international ones that disappeared with the ban. But at least for the moment there is a strange peace in the field.

So we feel that we can steal a few days to go lion watching. Soaring over the valley, we search for the big cats.

o o o

The lion glares into the eyes of his unlikely opponent — a Nile crocodile, whose pupils are as thin and cold as an ice pick. The giant jaws and ragged teeth of the two animals are locked into opposite ends of the glistening raw flesh of a dead waterbuck. The massive tail of the croc, plated with dragon armor, is coiled to launch the reptile forward in a flash. The leg and shoulder muscles of the lion ripple. Each of them has the strength and power of a small truck. Neither moves. The rest of the Serendipity Pride — three females, four large cubs, and another male — watch nearby, their ears turned back in apparent annoyance at this reptilian intrusion.

Mark pulls the plane around once more in a tight turn, as we watch the standoff on the beach below. We have never seen a lion and a crocodile challenge each other in this way. We fly back to the airstrip, grab our camping gear, and jump into the truck to get a closer look.

The Luangwa and its tributaries have one of the densest populations of crocodiles in Africa. During the rainy season many of the crocs migrate up the tributaries, returning to the Luangwa in the dry season when the streams dry up. A few crocs — such as this one on the sandbar of the Mwaleshi — stay in the shallow tributaries year round.

After the long drive from camp along the Mwaleshi, we park the truck under some trees that will make a good campsite, and hurry on foot to the riverbank.

"That's where they were, I'm sure." I point to a large sandbar on the other side of the Mwaleshi. From this side of the river there is no trace of a struggle between the lion and the crocodile, but we recognize the sandbar by its proximity to the bend in the river.

We wade into the clear, shallow water, heading for a smaller sandbar, covered in grass, that juts into the main channel. As our feet splash through the water, we both turn our heads this way and that, searching the sandy bottom for signs of a crocodile.

As I step up onto the sand, slightly ahead of Mark, I scan the deeper water on the other side of the bar. The river is faster here, and murkier, so it is difficult to see the bottom. But as I squint into the current, a curious pattern of reptilian scales takes shape, lying motionless only a yard from my feet. "Mark! He's right here!" I bend over, staring down at the coiled form.

Mark grabs my shoulder and pulls me back into the shallow current. "Delia, you can't just stand there watching a crocodile! He can fling himself out of there in a flash."

We back up, walk downstream for fifty yards, and cross to the large sandbar on the far shore where the skeleton of the waterbuck lies twisted on the beach. Kneeling, we study the lion tracks, carved deeply in the moist sand. They tell us that the lionesses grabbed the waterbuck as it stepped out of the tall grass of the floodplain and dragged it across the sandbar. Apparently, all of the lions fed for some time; we saw from the air that the females were quite full, and the tracks tell the same story. Then the lionesses left the carcass, probably chased off by the males. Soon after, the croc rushed in from the river and snatched one end of the kill. He was able to scavenge some of the spoils, but it was obvious that he did not drive the lions off; if he had, he would have dragged the carcass into the river or to his lair.

There are still many mysteries. Why did the other pride mates, the three lionesses and the male, lie basking in the sun while one male challenged the crocodile? Usually when other scavengers, such as spotted hyenas, vultures, or jackals, approach a fresh lion kill, the pride will chase them off. If the hyenas vastly outnumber the lions, the cats may retreat, but this was only one croc. His tough armor, lightning speed, and very capable jaws presumably give him license enough to feed with lions.

By watching Serendipity and her pride mates on earlier occasions, we have already learned that the Mwaleshi lions often hunt along the steep riverbanks where the prey species come down to drink. Thus their kills are accessible to the crocodiles. Furthermore, the Mwaleshi is so shallow that there are not many fish for the crocs to eat. To survive here they must be resourceful — even to the point of stealing meat from lions. Could scavenging from lions be a major source of food for these crocs? We will have to search for other lion-croc interactions to see how common such incidents are — and who usually wins.

The warm sands and cold bones yield no more clues, so we

wade back across the river, retracing our wide semicircle around the crocodile. After setting up camp on the banks of the Mwaleshi, we bathe in a very shallow area, taking more care than usual to check for predators.

Just after dawn, when the river looks like a ribbon of sunrise, the lions roar. We wiggle out of our sleeping bags, take a compass bearing on their position, and set off in that direction on foot. It is much more difficult to follow lions in Luangwa than it was in the Kalahari. In the desert, once we had a good compass bearing on lions, nothing stood in our way except rolling dunes; but here the land is crisscrossed with steep river cuts, dry streambeds, wet streambeds, rivers, lagoons, and eroded craters. At times we have spotted a lion only three hundred yards away and been unable to drive to it.

As we walk along the Mwaleshi, we come upon a herd of buffalo meandering toward the river. Creeping behind bushes and through the tall grass, we observe them from thirty yards away. Next we surprise a hippo wallowing in a shrinking lagoon. He whirls around and challenges Mark with a gaping mouth and squared-off teeth, while I scurry up a high bank to safety. We hear an elephant trumpeting and see the spoor of a mother and a tiny infant disappearing into the woodlands. There are no sounds or sights of man; it is a safari into old Africa.

The lions do not roar again, so we continue on the same bearing. When the noon sun is bleached to a blazing white, we see vultures circling over the river's edge and find them feeding on a dead bushbuck. We walk closer, our attention focused on the scrambling vultures.

A low growl erupts from the bushes. We swing around and two large male lions explode from the undergrowth, toward the river. Fifteen yards away they whirl around to face us. It is the two males from Serendipity Pride. We have made the mistake you cannot make with lions — we have crowded their space.

We stop in our tracks. They are in an aggressive stance, their heads and massive shoulders pulled to full height. One growls again through closed teeth, as they stare at us with piercing eyes.

There is nothing we can do but stand here; if we run, they will almost certainly charge. The male on the right raises his lips, exposing his canines in a snarl, then they both trot away, grunting loudly as they go. The entire encounter has lasted only ten seconds, but it has drained me. Mark and I lean against each other for a moment, and when we regain our legs, walk down to the beach to inspect the dead bushbuck. Although there are no signs of a crocodile, it is one more example of Mwaleshi carnivores making their kills near the river, where they could easily be confronted by crocs.

After several more long walks in search of lions, we decide to radio collar at least one of the Serendipity Pride members so that we can find them more easily. After dark we play tapes of lion and hyena vocalizations over loudspeakers near an open floodplain. Within thirty seconds, ten spotted hyenas are galloping toward the speakers, and ten minutes later the pride arrives. We want to dart Serendipity herself, but as they walk by the truck in single file, Mark has a better shot at one of the smaller females. He darts her. When the syringe stings her flank, she whirls around and trots toward the river. If she crosses the Mwaleshi before the drug takes effect, we will be unable to follow and collar her. Just at the water's edge she sways and stumbles, and finally succumbs.

As I watch the other lions with the spotlight, Mark walks up to her and nudges her gently with his foot to ensure that she is properly sedated, then I move the truck closer. As we begin to collar her, the other lions lie watching in a rough semicircle forty yards away. Shining the spotlight around every few minutes, I keep an eye on them; but the spotlight bulb burns out. Unbelievably, the backup spotlight also blows. We have to keep tabs on the curious, undarted lions with the weak beam of our flashlight as we collar, ear tag, and weigh the lioness.

Finally we finish and name her Kora, after the beloved Kenya bush country of George Adamson, who was murdered by Somali ivory poachers. These poachers cross into Kenya, shoot elephants, and smuggle the tusks back to Somalia, which is heavily involved in the illegal exportation of ivory. Adamson had attempted for years to defend the Kora Reserve against these pirates, but eventually they shot him to death and he became yet another victim of the ivory trade.

The next morning, by listening to Kora's signal, we find that the pride has crossed the river. Carrying the antenna, radio receiver, and rifle, we wade across the Mwaleshi, following the beep-beep-beep of her collar. On the other side, the grass is so high that even when we part it with our arms we still can see only two feet ahead. The needle on the amp meter goes into the red, indicating that we are very close to the lions, but all we can see are grass stems and blue sky. We listen for the sounds of lions walking in the grass. Nothing. Mark motions me forward, but I point behind us, suggesting that we retreat. Mark shakes his head, and we walk on slowly, parting the grass, trying to see ahead, listening.

The signal remains strong and constant as we go forward. The lions must be staying just ahead of us, moving in the same direction at the same speed. Quiet as we try to be, we know they can hear us coming. Again I suggest a retreat. Whispering, I say that we can't see them in this tall grass anyway, and we might stumble upon them unexpectedly and frighten them. I don't mention that I am already scared silly, and that this scheme is just plain foolish. "Just a few more minutes," Mark whispers. We stalk on through the grass for another ten minutes before giving up and heading back to our little camp on the Mwaleshi.

The next day we walk the beach, following the lion tracks and scanning for vultures, but we cannot find the pride or pick up the signal. Tomorrow we must return to Marula-Puku to continue the antipoaching and village programs. But as soon as time allows, we will return with the airplane, find the radio-collared lioness, and observe the pride in more detail.

As we are breaking camp at dawn, we see a swarm of vultures across the river on the same sandbar where we first saw Serendipity Pride and the croc. Pulling and tearing on a puku carcass, the vultures glare with beady eyes as we wade the river. When we step onto the sandbar, they lift off in a flurry of flapping wings.

The vultures have not been here long, and the story of the kill is still etched in the sand: the lions killed their prey on the sandbar, and once again the crocodile rushed from the river to grab a share.

"This croc's really got something going," Mark says, as he inspects the remains of the puku.

"And there he is." I point to the river's edge, where the croc is lying in shallow water, his massive back rising above the surface, his flat head held up, mouth open, exposing a row of jagged, uneven teeth. He looks incongruously as if he is grinning. Mark steps slowly toward him, and I follow a few steps behind. The croc doesn't budge at our approach; he obviously owns this beach. When Mark is ten yards away, the croc hisses and snaps, warning him not to come any closer. And I remind Mark of his words: "You can't just stand there watching a crocodile."

As we turn to leave, we see a cyclone of black smoke rising in the south. The poachers are back in the park. Our short days of lion watching are over.

o o o

The wide, unfamiliar waters of the Mutinondo River spread before me. The only other time I have seen this river, it was a raging torrent, sweeping past with driftwood on its bow. Now its clear waters lap gently at grassy shores and its current whispers quietly over polished stones.

"Well, it looks okay to me. What do you think, Marie?"

Marie Hill, a Texan, and her husband, Harvey, live in Mpika, where he is the representative for the Canadian Wheat Project. Marie has volunteered to coordinate our conservation education program, which has grown too large for Mark and me to handle from camp.

"Yeah, it looks okay to me, too," Marie drawls.

"Let's go for it." I ease the front tires into the river and start across. We're on our way to Nabwalya, one of the picturesque villages in the Luangwa Valley, to begin our village and school programs. Many of the poaching expeditions into the park begin in Nabwalya; it is essential to win the people there over to our side.

To get this far I have had to drive up the scarp to Mpika to pick up Marie, and down the scarp on another track south of the park. Nabwalya is completely cut off during the rainy season, and not many trucks reach the village at any time of year. My Toyota is overflowing with medicines for the clinic, materials for the school, emergency protein rations for the hungry, our gear — an array of duffel bags, sleeping bags, camera cases, food boxes — and a watermelon for the chief. Balancing on top of this load are our educational assistant, the village medical officer, a schoolteacher, and the Nabwalya mail carrier, who usually has to walk for four days to reach the village from Mpika.

The river is shallow and easy to ford, and although the rest of the track is grueling at times, we make it to the village by four in the afternoon. The first thing we must do is pay our respects to the chief, so we send a runner ahead to announce our arrival. After we drive across an abandoned field, we find him and his headmen waiting for us in his n'saka. Nabwalya is young for a chief, alert, articulate, and progressive. He welcomes us warmly in perfect English, and we pass out our gifts — *International Wildlife* magazines and the watermelon. After a few moments of polite greetings, we ask for permission to begin our programs in his village, and he says that he has been waiting for us.

"Of course, you will stay in the 'guest palace,' " the chief says. "We have made a very nice guest palace for the tourists when they come. My headman will show you the way."

The new guest area is indeed a palace: two large bungalows, a n'saka, and a latrine, all made of woven grass and reeds. The roofs, doors, and windows of each structure have been trimmed with decorative spirals of grass. The charming encampment is perched on the high bank of the wide Munyamadzi River, just outside the village. A group of hippos have already begun their night song as we start to unpack. I am enchanted with the guest palace, but a little worried about how soon Chief Nabwalya expects the tourists to start coming.

The first camper to rise is usually rewarded, and the next morning is no exception. As I stoke last night's coals at 4:45 A.M., I hear the swish, swish of hippo feet in shallow water. Below me, just as the river turns a bright orange from dawn, a hippo strolls past, silhouetted against the shimmering water. In the faint light I can see the outlines of thatched huts and the smoke from a few cooking fires. It is a rare moment of people and wildlife together in harmony.

It is also the last moment of peace we have for four days. Marie, Mukuka (our assistant), and I present our conservation slide show to one hundred fifty villagers packed into the mud-brick school-house. At our prompting, Chief Nabwalya explains to his people that many of their problems will be solved if the village can make money from tourism, but it will happen only if the poachers stop shooting the wildlife in North Luangwa.

With many of the curious villagers in tow, we deliver medicines to the clinic and talk with the medical officer about his problems. The first thing we notice is that all the patients are lying on reed mats on the floor, while the rusty hospital beds are stacked in a corner draped with spiderwebs. The medic tells us that poachers have stolen the bolts from the beds to make bullets. Never missing a chance to deliver our message, we point out to the patients that the poachers are their enemies and that we will get more bolts for the beds.

Next we inspect the airstrip, which is under construction by laborers we had hired earlier. When it is complete, the Flying Doctor can land and the village will not be so isolated during the rains.

Sunlight beams through holes in the walls of the schoolroom. Standing at the front, Marie and I encourage the children to draw posters asking the poachers not to kill the animals near their village. We wait for the children to color on the white paper we've given them, but unlike the students in the other villages, the Nabwalya children only stare at the page.

"Why aren't they drawing?" I whisper to the teacher.

"They have never had their own sheet of clean, white paper," he whispers back. "They are afraid to spoil it."

After we promise the children that we will bring more paper for them, they slowly begin to draw. A little girl sketches a hippo with ten babies under the caption, "Please Mr. Poacher Do Not Shoot the Hippo She Has Many Children." Later, walking along the well-worn footpaths, we and the children post their pictures throughout the village on trees and stumps.

On visiting the game guards of Nabwalya, we discover that they have never mounted a patrol in North Luangwa because they would have to cross several rivers with many crocodiles. When I point out that the poachers cross the rivers, they tell me that crocs do not attack criminals. They agree to patrol the park if we buy a small banana boat for them. Before leaving their camp, we give out T-shirts to all the game guards' children.

Steaming campfire coffee and hippos silhouetted against the dawn begin the last morning. The sun is especially welcome, because the village drummers have celebrated a wedding all night and only now end their rowdy beat. Exhausted, Marie, Mukuka, and I cook toast and oatmeal over the fire. We still face breaking camp and driving up the scarp. As I struggle toward the truck with a large food box, I hear a familiar sound drifting in from the distance. Our plane. None of the others have heard it.

I rush to the riverbank and wait. I know Mark; he will soar along the river course, just above the water and below treetop level, until he finds me. He knows me, too, and has guessed that I am camped somewhere along the river.

In seconds the plane appears just above the water, almost level with our grass palace. Marie and I wave frantically. Mark holds a note from the window, signaling that he will drop it on the next pass. I run to a clearing behind the camp and watch him circle back, figuring it is a shopping list for Mpika or some other errand he wants me to run before returning to camp. As the plane soars past, Mark flings the note, tied to string and a rock, out the window. I tear it open and read: "Greetings, my love. Come home to me. You have been away too long. I have a special surprise for you. Love, M." True, there is a short shopping list below, but the love part comes first. I hold the paper high in the air, signaling to Mark that I have received his message. He waggles his wings and flies south toward camp. "It's a love letter," I say to Marie. "He flew all this way to bring me a love letter. I've got to get back to camp!"

Two days of hard scarp driving makes my shoulder muscles

burn. But I pull into camp late in the afternoon of the second day. Mark welcomes me with a bear hug.

"Where's my surprise?" I smile.

Mark can barely contain himself. He pulls me along to the porch of our bedroom cottage, tells me to sit down, and hands me a glass of cool wine. "Okay, we wait here. The surprise is coming." As we chat quietly, Mark keeps glancing at his watch and then toward the tall grass bordering camp.

Minutes later, he touches my arm and motions for me to stop talking. He points behind me to the edge of camp. "There, there is your surprise."

I turn and see an elephant — Survivor — only forty yards away. He lifts his trunk and holds it high, taking in the amalgamation of camp scents. I squeeze Mark's hand, but otherwise we are still. "Just wait," Mark whispers, "just wait."

Survivor takes one step forward, then stops and lifts his trunk again. He raises his right front foot as if to step, then pauses and swings it back and forth. For long minutes he dangles his foot and lifts his trunk, then quite suddenly walks toward us with no further hesitation. He marches right into camp and begins feeding on the marula fruits behind the office cottage. Twenty-five yards from us, he feels along the ground with his dexterous trunk, finds a fruit, pops it into his mouth, and chews with loud slurping noises. Now and then he gently swings his entire head in our direction, but otherwise he does us the honor of ignoring us completely. He feeds for thirty minutes; just before dark, he walks away quietly along the same path he had followed into camp.

Mark tells me excitedly that Survivor has been coming into camp every evening since I have been gone. That this kind of acceptance can happen so soon, in the midst of such a slaughter of his kind by our kind, is all the proof we need that our project may work. It is stimulus enough to keep us going.

Survivor continues to come to camp every day, not just in the evening but anytime. Mornings he ambles up to the kitchen boma and feeds ten yards from the campfire. He forages between the office and bedroom cottages, and once he walks within four yards of us, as we stand quietly on the porch of the bedroom cottage — so close he could touch us with his trunk. No longer do we have to freeze in his presence. There is an unwritten truce — as long as we move quietly and slowly around camp, he pays us no mind. Simbeye, Mwamba, and Kasokola are as taken with Survivor as we are, and often I see them standing quietly at the workshop, watching him feed.

The truce is broken only once. One morning Mark hurries down the path toward the office cottage, his head down. As he steps around the cottage, he looks up to see Survivor's knee only six yards from him. The elephant, caught by surprise, draws up his trunk and flaps his ears, making loud blowing noises as he steps backward. Pivoting around, he runs twenty yards before swinging to face Mark again. Mark resists the urge to flee in the opposite direction and stands quietly. After a few minutes of ear flapping, Survivor calms down and feeds again.

o o o

The annual inspection of the airplane is due, so Mark has to fly to Johannesburg to have it serviced. Unfortunately, he will be away for more than three weeks. I stay behind to continue our work and to watch Survivor. Occasionally the guys and I see him and his four companions roaming the hills between the airstrip and camp. All of them have become accustomed to us and do not run away — but only Survivor comes into camp.

Elephants can move through the bush as quietly as kittens, but when they feed, they make a noisy racket as they strip leaves from a branch or topple small trees. Whether I'm working at the solar-powered computer, building the fire, or reading, I know where Survivor is by his loud slurps. At night he drifts through the sleeping camp like a large moon shadow. Lying in bed, I am lulled to sleep by the stirring wind, his soft footfalls, and the rustling leaves — the song of Survivor.

One night as he forages in front of the cottage, I cannot sleep, or perhaps do not want to. I get up and ease the door open. The half-moon winks through the marula leaves as Survivor feeds beneath the tree ten yards from where I stand. Inching forward, watching his every move, I step onto the stone veranda and slowly sit on the doorstep. Gracefully he turns to face me, lifting his trunk to take in my scent. He immediately relaxes, lowering his trunk and sniffing loudly for another fruit.

Ten yards from an elephant. Sitting down and looking up at an elephant. He covers the world. Even in this soft light I can see the deep wrinkles and folds in his skin. They look like the craters and valleys of the moon.

Slowly rocking back and forth, he picks up the fruits, sometimes extending his trunk in my direction. Gradually he makes his way around the corner of the cottage, until I am sitting in the moonlight watching his huge backside and his tail. Even that is enough. When at last he disappears into the shadows, I return to bed and sleep.

It is, of course, the marula fruits that keep him coming. In early May thousands of them carpet camp, releasing a sweet scent into the soft air. But by late June most of them have been eaten by Survivor and the Lillian lovebirds. Once the fruits are finished, he will migrate to the scarp mountains. I wonder, in spite of his name, if he will survive another season. Every morning I poke around under the bushes, under the solar panels, in all the hard-to-reach places, pulling the fruits out into the open where he can find them.

Survivor is already coming into camp less often, maybe every other day instead of daily. Only rarely do I see him and the four others feeding on the tall grass and short bushes of the hillsides.

Now the marula fruits in camp are nearly gone. At night I sit at the window watching the moon cast eerie shapes across the grass. But the large moon shadow and the soft song for which I wait come no more.

16

One Tusk

MARK

> I am the wind.
> I am legend.
> I am history.
> I come and go. My tracks
> are washed away in certain places.
> I am the one who wanders, the one
> who speaks, the one who watches . . .
> the one who teaches, the one who goes . . .
> the one who weeps, the one who knows . . .
> who knows the wilderness.
>
> — PAULA GUNN ALLEN

o o o

WHERE THE LUBONGA meets the scarp, One Tusk carefully picks her way down a steep ravine, stepping gingerly over sharp gravel on the path, her massive body swaying rhythmically, almost as though she is on tiptoe. Her single tusk, more than three feet long, seems to touch the side of the stream cut as she looks back at her family. They include Misty, her eighteen-year-old daughter, and Mandy and Marula, two younger adult females. One infant and a three-year-old calf follow closely, watching their footing as they maneuver down the slope.

The matriarch has passed this way many times before, as her ancestors did. Each time she pauses, as she did earlier today, to smell and fondle with her trunk the bones of the recent dead. But now instead of thousands of elephants, only small, isolated groups move silently, nervously, through the dying forests toward the Lubonga River.

The winter sun, dull and bloated from the smoke of dry-season wildfires, sags into the distant hills. One Tusk will take her small family to drink at the river only after dark, when the poachers have camped for the night.

Staccato bursts of gunfire thunder through the canyon. Several bullets slap One Tusk in quick succession. She staggers back, raising her trunk and screaming a warning, eyes white with fear, blood streaming from small holes in her neck and shoulder. She screams again, a pink froth foaming from her mouth. Three more blasts shatter her skull. Sinking to her knees, as though in prayer, she slides in that position all the way to the bottom of the bank. The calves run to their fallen matriarch, wailing in confusion. Their mothers turn to face the gunmen.

One of the shooters steps boldly to the top of the bank in full view of the terrified elephants. Shrieking, ears back, trunks raised, Misty, Mandy, and Marula charge up the slope toward him. More shots ring out. Two puffs of dust explode from Mandy's chest. She screams, sits on her haunches, then rolls back down the hill in a tangle of legs, trunk, and tail. Misty and Marula break off their charge and stampede out of the ravine, the calves huddled close to their flanks. They run for more than a mile, to a meadow below the scarp where there are dense thickets, and a stream from which to drink. Without One Tusk's leadership, they mill about in utter confusion.

The beleaguered elephants press together, turning circles, their trunks raised, sniffing the air. Over and over they entwine their trunks and reach out to touch faces, apparently reassuring themselves that the others are still there, and are all right. The mothers reach down with their trunks, pressing the calf and infant close to their sides, caressing their faces, necks, and shoulders.

After they have calmed themselves, Misty, Marula, and their young walk farther east, away from the point of attack. Then they turn back and spend the night huddled together just a few hundred yards from where One Tusk and Mandy were shot.

The next morning, rumbling loudly and holding their trunks high, their temporal glands streaming, they walk to where One Tusk and Mandy lie sprawled in a great reddish-brown stain made by their body fluids. The faces, feet, tails, and trunks have been hacked off the slain elephants, and they are buried beneath a seething pile of hissing vultures, which have fouled the gray and bloating bodies with white streaks of excrement. Swarms of maggot flies have already begun laying their eggs in the rotting flesh.

Leading the others, hesitating after each step, her trunk raised, flapping her ears, Misty cautiously approaches the mutilated corpse that just hours ago was One Tusk. She and the others assemble around it, their trunks extended, smelling every inch. Then Misty nudges it with her front foot and tusks. The others join in, as if willing their slain matriarch to get up and live again. After a long while Misty walks to the severed trunk, picks it up with her own, holds it for a moment, and then puts it down, her cheeks wet from her weeping temporal glands.

After smelling and fondling their dead for nearly two hours, Misty and Marula begin ripping up tufts of grass, breaking twigs from trees, and scuffing up piles of gravel with their feet. The younger elephants join in as they cover the corpses with this debris. The burial takes the rest of the morning, and when they are finished, the elephants stand quietly, their trunks hanging down for several minutes. Then Marula leads them away. Misty returns to her slain mother, puts her trunk on her back, and stands for nearly two minutes. Finally she turns to follow the others.

She catches up to the little group and assumes the lead. Walking at a determined pace, they all head northwest toward the tall peak of Molombwe Hill, which is set behind the chain of hills that forms the leading edge of the scarp. Below it is Elephant's Playground, where they last saw Long Ear and her family unit. Rumbling constantly, temporal glands streaming, they follow the paths used by their relatives as they cross the hills and vales along the foot of the Muchinga. Often they stop to listen, and to smell, circling with their trunks raised high, testing the breeze. The elephants have hardly fed for two days, and the two youngsters are especially hungry and tired.

On the morning of the third day Misty, Marula, and the others are nearing Elephant's Playground when they hear a rumble ahead. Running along the slope at the edge of the swale, they see Long Ear's group watching them, trunks raised. Misty runs faster, as Long Ear and her two young daughters rush to meet them. The trumpeting, rumbling, and clacking together of their tusks echo to the hills as the two groups entwine in greeting. Pushing together, milling in a tight compaction of gray bodies, over and over they touch one another's faces with their trunks. Long Ear keeps looking east, as though expecting One Tusk and Mandy to appear. Misty walks into Long Ear's side and stands there, her face buried in the loose and wrinkled skin.

Eventually the elephants calm down, but they continue rumbling softly and touching one another. Both groups have lost their matriarch — and with her, access to the immense wisdom acquired from her ancestors and from her decades of roaming the valley. Young, inexperienced, relentlessly pursued, these orphans of the poaching war will stay together, raising their young as best they can. They move off now, deeper into the hills, touching often.

o o o

Harassing the poachers with the plane is no longer working. I had flown over One Tusk and her family just an hour before she and Mandy were killed. The poachers certainly saw the plane, but it didn't frighten them anymore. They have figured out that I can't land except on an airstrip or a sandbar. The next morning, drawn by smoke from a fire the killers set on their way out of the park, I fly again and find the dead One Tusk and Mandy, their family milling around them.

Flying a patrol along the Mwaleshi four days later, I see two elephants grazing near the river. On my way back to camp an hour and a half later, I discover their tuskless carcasses. Landing along the river, I hike to the spot, where I record their age and sex and look for any evidence — a shirt, a boot track, rifle cartridges, a piece of camping gear — that might lead us to the poachers. Then, afraid they may still be in the area and attack the plane, I run back and take off for camp.

I am two hours overdue by the time I get back to Marula-Puku. Delia meets me at the workshop, her face pale with worry. "Mark, where have you been? I was just about to call out a search."

"I've been out counting dead elephants, where else!" Slamming the door of the truck, I stomp off to the kitchen for a cup of black coffee. For some time I have been living out of my coffee cup, drinking a brew so strong it is like a thin syrup. The caffeine gives me the kick and the courage to do what I have to do, but increasingly it is running my brain and my mouth. My fuse is very short.

Poachers have invaded large areas of the park, including the sanctuary around our camp, which until now I have been able to defend with the airplane. The region has become a war zone. From the air one morning I discover six dead elephants, faces chopped off, tusks removed, on the Lubonga near its confluence with Mwaleshi. Two days later a freshly set bush fire leads me to another two carcasses on the Lufwashi near Chinchendu Hill. The next day three more are slaughtered along the Mwaleshi River at the north side of Chinchendu. And then three more north of Marula-Puku — killed by Patrick Mubuka, the officer in charge of the Nsansamina game scout camp.

I am flying night and day, spotting poachers, diving on them with the plane, and airlifting scouts when they will come. But the killing is totally out of control and escalating. Delia and I cannot go on like this for much longer, and neither can the elephants. But we have at least solved a mystery: we have found the fabled elephant graveyard. It is Zambia.

On long night flights, droning along in the darkness while searching out poachers' fires, I find myself staring into the plane's black windscreen, the faint red glow from the instruments reflecting back at me, trying to figure out why the poaching has so suddenly escalated. Even though it is not as bad as it was before the ivory ban was enacted, it has exploded recently and I don't understand why.

One morning a few weeks after the killing of One Tusk and Mandy, I am standing next to our Mog at the Mpika marketplace. Local people are coming and going to the cubicles and stands that sell potatoes, mealie-meal, cabbages, dried fish, and black-market cooking oil, sugar, and tobacco. Kasokola is inside doing the buying while I guard the truck. I feel a little uneasy, as though every poacher in the district is watching me. After about twenty minutes a Zambian man dressed in light gray slacks, an argyle knit shirt, leather shoes, and sunglasses slopes toward me across the dry and dusty ground.

"Mapalanye," I greet him in Chibemba.

"Good morning. How are you?" He is slightly unsteady and smells strongly of beer.

He studies the Frankfurt Zoological Society logo and the North Luangwa Conservation Project lettering on the side of the Mog. Aloud he misreads, " 'North Luangwa Construction Project.' So you work in construction. Where?"

"Down in the valley." I wave east, toward Luangwa. "I'm in road construction."

"I see." Pause. "You want to buy something, maybe?"

"Depends what you're selling."

"Aaaanything! Anything you want." He holds out his arms. "You want ivories, I have ivories; you want lion skins, I got lion skins; elephant foots and tails; zebras, leopards, anything." He leans close to me, lowering his voice as some women come by selling a basketful of dried fish. "I even got rhino horns. You interested?"

"Well, I don't know. I'd have to see the goods first. Where are you getting all this stuff?"

"I know people with a safari company. My merchandise mostly comes from Fulaza Village, near North Luangwa. They organize the hunters that side, some from Mpika, some from Fulaza, then transport the ivory and skins to Mpika."

We step aside as Kasokola and another man lug a bag of beans by its ears to the back of the truck and throw it in. "How much do you have to pay a hunter to shoot an elephant?"

"For now, not so much. Only about a thousand kwachas." That's a little over ten dollars — for an elephant.

He tells me that a good hunter can earn as much as fifty thousand kwachas in three weeks in the bush, and up to half a million in a year. The illegal ivory is transported to Lusaka in army trucks or in civilian vehicles, hidden in spare tires, bags of mealie-meal, drums, or buried at the bottom of heavy loads. In Lusaka the ivory takes various routes out of the country. Some of the tusks are cut up in dozens of illegal "chop shops" and smuggled in small pieces to Swaziland aboard Swazi Airlines. Or a government contact is paid to launder it with official documentation, and it is shipped with legal ivory through Botswana and Zimbabwe to South Africa. From there some of it is carved and sold as crafts; the rest is marketed to dealers in China, North Korea, and other non-African countries that refuse to observe the ban on trading ivory.

"This ban is troublesome," the dealer complains. "For now it has made ivory too cheap. We survive by selling to dealers in countries still trading." He refers, of course, to South Africa, Zimbabwe, Botswana, Mozambique, Malawi, Angola, and Zambia, which refuse to observe the moratorium. "But we hope this will not last. Countries will vote again soon. So we are shooting elephants and burying ivories, waiting for America, Europe, and Japan to refuse this foolish ban." He leans against the Mog, his hands in his pockets.

"What if all countries stop trading ivory?" I ask.

"Ah, but then there will be no market at all and my business will be finished. I will have to go farming. Come, Big Man. Let us do some business. What you want and how much?"

"Right. Let's see some of your goods." Gary Simutendu spells his name for me and tells me how to get to his house. I promise to stop by in a day or two. Instead, I head straight for Bornface Mulenga, the warden. He tells me Simutendu is no longer dealing, that I should just forget what he said. Instead, I fly to Lusaka to talk to Paul Russell and Norbert Mumba at the Anticorruption Commission. Months later they will break up several dozen illegal ivory-processing factories in Lusaka and indict a Chinese diplomat and three senior administrative officers in the Department of National Parks for smuggling ivory to Swaziland and China.

Simutendu, drunk as he was, told me all I need to know about how to save the elephants of Luangwa: it will be impossible without a total, long-term, worldwide ban on trading ivory and other elephant parts. We will do everything we can to convince Zambia and other nations to join the ban, but until they do it will be up to us to get rid of poachers like Chikilinti.

One of our newly recruited agents is an ex-poacher who has hunted with Chikilinti many times. Several days after my encounter with Gary Simutendu, we set up this informant with money for beer, four rounds of shotgun ammunition, and the story that he has been jumped by scouts while on his way into the park to poach. We send him to Mpika. Finding Chikilinti at a bar in Tazara, he buys him beer after beer until he is drunk. Then he slips away to the warden's office, asking that he send scouts to arrest the poacher. Two scouts hurry to Chikilinti — and warn him that he is about to be apprehended. Immediately he melts into the crowded street, and we have lost our best chance at nailing the godfather of all poachers.

o o o

Perched in our canvas chairs high above the murmuring Lubonga, we watch the thunderheads tumbling and growling in the bruised sky above the stony rubble of the riverbed. The butter-colored light of the setting sun comes and goes between the clouds, bringing out the brilliant green of the young grasses springing up in the sandbars along the floodplain. The puku are sprinkled like flakes of red cinnamon through the grasses, and now and then a flock of snow-white egrets or a yellow-billed stork soars past, flowing with the river course.

The color seeps from the sky, and dusk settles over the riverbed community. A flock of guinea fowl rails plaintively to our right; a fish jumps in the pool below our feet, and the puku and waterbuck take tentative steps toward the river for a drink. Pennant-winged nightjars, wispy phantoms of the twilight, trill as they sail above and around us, so close that their wings almost touch our faces. The darkness deepens, and the soft forms of the antelope slowly dissolve into the gray stones across the water.

An elephant's life is worth about ten dollars.

We can talk of nothing else. So mostly we don't talk. There is a time to quit, and this is it. We must give up; giving up is survival. Delia leans her head back, looking up at the sky, and when she speaks her voice is far away and very tired. "I can't remember when we last looked for shooting stars. Let's go home for a rest," she pleads. And I agree, it's time to get away for a while.

Several days later, leaving the guys in charge of Marula-Puku, we drive to Lusaka. Two hours before our flight to the States, we get a radio message from Marula-Puku: gunshots have been heard all around. Poachers are killing elephants within a few hundred yards of camp. The director of National Parks agrees to send paramilitary scouts to protect the camp while we are away.

In Atlanta, only two weeks after arriving, I take an urgent phone call from Marie Hill in Mpika. Late last night, she tells me, the home of another of our informants was raked with AK fire; some nights before that, still others were beaten; and Mathews Phiri, a trusted scout whom we brought from Livingstone in southern Zambia to work undercover in Mpika, is sick in the hospital after being poisoned. We tell Marie to have Musakanya call in all the undercover agents until we get back.

We needed to get away, but it seems eight thousand miles is not far enough.

A few days later Delia and I are in a lounge at Atlanta's Hartfield Airport, waiting for a flight to Charlottesville, Virginia. Returning to our table with her drink, Delia staggers and clutches her chest. Before I can jump up to help her she collapses into her chair, spilling her drink over the table, her face twisted in pain, her breathing ragged and shallow. I run from the lounge to the boarding gate nearby, and the attendants call for the airport's paramedic team. Within two minutes they are setting up a portable electrocardiogram machine in the lounge and stabilizing Delia so that we can get her to a specialist.

In the cardiologist's office the doctor warns her: "Your body is trying to tell you something. I can prescribe some drugs — beta blockers — but the main thing is that you're going to have to change your lifestyle in order to minimize stress." Delia looks out the window and laughs.

17
The Eye of the Storm

MARK

The storm clouds come rolling in, and night is down upon
us with a poison wind . . . We have all but lost.
— JIM BARNES

o o o

"ENCOUNTERED A GROUP of one hundred plus poachers moving down Mwaleshi. Group had twelve weapons including automatic military rifles. Surrounded our porters, but released them
later. Appealed to Chanjuzi [scouts] for help but they were inca-
pable of raising more than one weapon with four rounds and no
mealie-meal. Please assist with antipoaching. As Mulandashi River
now unsafe for walks must fly with you to find alternative areas so
won't have to cancel rest of safari season . . ."

The radio crackles with this urgent telex message from Iain
MacDonald. Months earlier he set up a reed and thatch camp on
a lagoon near the Mulandashi-Luangwa confluence and organized
photographic walking safaris, the first full-time tour operation in
North Luangwa. We are anxious for him to succeed.

After three weeks in the United States we are back at Marula-
Puku. In spite of our talk about quitting, we have never seriously
considered it. We cannot abandon the park and its elephants. The
problem with Delia's heart is a faulty valve, its performance made
more inefficient by stress; but as long as poachers are attacking the
park and game scouts refuse to work, our lives will never be peaceful. So we have come back armed with beta blockers and a new
determination.

At the airstrip on the night of Iain's message, Kasokola, two
guards, and I set out our milk-tin flare pots. We have just lit the
last one when a big thunderhead begins grumbling southwest off

the airstrip, between us and Chinchendu Hill. Kasokola and I wait in the plane for forty-five minutes, hoping the storm will move on, but it just sits there. After another fifteen minutes, the thunder and lightning start to fade and I decide to take off.

I taxi to the west end of the strip, run my power checks, take a last look at the sky — and decide to wait again. I can see the black cancroid mass of a broader storm system moving in with the thunderhead. Although the sky to the north is generally clear, I don't want to take off and then not be able to see the horizon properly, especially if the storm moves north again. My plane's vacuum pump is broken, so the artificial horizon and gyrocompass are useless. My radar altimeter, which would give my exact height above ground, doesn't work either; and with the changing barometric pressure associated with the storm, my pressure altimeter will be very unreliable. Some days ago I managed at last to fix my airspeed indicator: Cedric, a crafty mouse who was living in the plane, had chewed through the line from the Pitot head to the pressure canister under the instrument panel.

It is still almost pitch black, but I think I see dawn growing under the sleepy eyelid of the eastern sky. "Do you see it, Kasokola?" Yes, he thinks he does. I'm going for it. If we wait any longer we will have a hard time spotting Chikilinti's fires after daybreak. For this is Chikilinti, all right; I know by now that any poaching group this large will have been organized by him.

Almost as soon as Zulu Sierra's wheels leave the ground, I'm sure I have made a mistake. There is no sign of dawn and I can barely make out the horizon to the east. All other corners of the sky are pitch black. If the storm envelops us, I will be forced to fly on instruments, and the ones I need most are not working. I am flying out of the realm of reasonable risk and into the realm of stupidity. As we pass over camp I turn the plane around, intending to get back on the ground as quickly as possible. The new beacon beams at the east end of the strip, but the west end has already been swallowed up by dark clouds and heavy rain. The downpour has drowned six of the nine flares, leaving the south side of the runway in total darkness and only three flares and the beacon on the north side.

"Mark, are you there?" Delia is calling over the radio. "It's really blowing hard down here. Do you read me?"

"Roger, Boo, it looks a little nasty to the south, so I'm coming down right now. Over."

But for some reason Delia cannot hear my transmission. "Mark! Do you read me? *Do you read me, Mark?*"

"Roger, I am coming down right away."

"Mark, I repeat, there's a bad storm moving in. *Do you read me?*"

"No time to talk now, love. I'll see you when I get down."

"Mark, I cannot read you. Are you okay? Answer me, please . . . Oh God . . ."

A giant fist of wind slams into the plane, flipping it on its right side. Rolling the controls to the left, I try to steady it. Maybe I should buzz camp, to let Delia know I haven't gone down. No time. Two, maybe three, minutes is all I have to get back on the ground before the entire strip is buried in cloud and rain. I shove the control wheel forward, diving toward the runway. At the same time I reach behind my head to turn up the red instrument panel light so I can see the altimeter unwind. Whatever else, I must not go below 2300 hundred feet. If I do, we're dead; for I have previously calculated the airstrip's elevation at 2296 — where soft sky turns to hard ground.

Descending through 2900 feet, I turn on my landing light. But the rain pitches the beam back at my face, a white sheet that totally blinds me. I quickly switch off the light and steepen my dive for the friendly, mesmerizing beacon, pulsing like a firefly at the end of the strip.

Then the spluttering flares and the beacon are gone. The storm is shouldering its way up the airstrip toward me. Pulling the nose of the plane up a bit, I slow our rate of descent while trying to find the beacon or flare-path. I glance at my altimeter again. Its white needle is nearly touching 2300! My God, we're almost on the ground! But where is it? I can't see a thing.

Rain and hail hammer the fuselage like buckshot, and my windshield is black. I fumble for the landing light switch. Suddenly the beacon reappears out of a paunch of cloud, beaming cheerily atop its twelve-foot pole — in front of my right wing. I throw the control wheel over, lifting the wing, and pull it back to bring us out of the dive. With my finger I stab the switch. The powerful beam of light cuts through the darkness.

And there is a tree . . . right in front of me . . . not fifty yards away. We are below its crown, within ten feet of the ground. Without a lighted flare path I have misjudged not only our height but also our alignment with the airstrip. We are landing crabwise at about a thirty-degree angle to the strip. Traveling at a speed of seventy knots, in exactly 1.27 seconds we are going to hit the tree.

I ease the controls back and shove the throttle forward. The engine roars and the plane surges, as if someone has booted her in the backside saying, "Come on, baby; now is the time to show us you can fly." She points her nose bravely at the storm, and as she claws her way up, her landing light lifts. The tree disappears from view.

I grit my teeth, listening to the tearing sound as some part of the plane rips through the tree. But we are still flying. Later I will find shrubbery caught in the left main wheel, and green stains on the undercarriage.

Now I can see nothing, and Zulu Sierra is bobbing and jerking. Wrestling with the control wheel, I watch the turn and slip indicator, trying to keep its miniature wings level so that we won't spin out and crash. This instrument was never meant for flying blind, but it's all I've got. "Keep climbing! And don't overcontrol! Get away from the ground!" my mind shouts at me.

The racket of the rain drumming on the skin of the plane becomes a dull roar. Through the windows all I can see is the blackest black I've ever seen: no flares, no beacons, no moon or stars. But minutes earlier the sky to the north was clear. Careful not to overcontrol and spin out, I push right rudder, ease on a little bank, and begin a very gradual northward turn from my southwest heading.

For several minutes I fly through air like lumpy ink, feeling my way out of the black belly of the storm. Finally I see a faint amber glow in the windshield — the beacon on the airstrip. Damn! I'm diving into the ground again! No . . . I force myself to believe my altimeter. It shows 4000 feet and climbing. Anyway, flying north, the light should be behind us. Looking around, I see the beacon through the right rear window. It is reflecting in the windshield. We have flown out of the storm. A string of yellow flares is growing along each side of the airstrip. The guys, God bless them, are busy trying to get us home.

I circle in the clear air for several minutes while the storm drenches the runway and moves on. Then I turn back and land on the soggy strip, the plane's main wheels spraying the underside of its wings with mud and water. I taxi Zulu Sierra to her boma, switch off, and slide from my seat. As the guys tie her down and pick up the flares, I stand alone on the brow of the hill, sucking in the sweet, wet air, listening to the last grumbles of the dying storm. I've had enough; finding Chikilinti can wait for better weather. Before heading down to camp, I give each of the guards a pat on the back and a fat bonus.

Delia is sitting on the bed of our stone and thatch cottage, crying in the dark. I switch on the solar light and try to hold her. But she pushes me away.

"Mark, I thought you were dead! I actually thought you were dead! There were no trucks here and my radio wasn't working, so there was nothing I could do about it. And this happens night after night! Do you know how that feels? Do you care?"

Turning to the window, I part the curtains and watch the heavy clouds turn silver above the rising sun. "I'm sorry, Boo. I just don't know what else to do. You know my flying is the only thing standing between the poachers and the elephants."

"Yes, but I'm not sure this is worth dying for anymore. If your dying would change anything, then maybe it would be. But it won't. And I don't want you to die for nothing! I want to stop the poaching as much as you do, but you've crossed over the line, and I can't go on like this. I just can't. I know you can't change; you're doing what you believe is right, and I respect you for that. You'll keep after the poachers until you either drive them out or fly into the ground trying. But I cannot sit here night after night, day after day, waiting to see if you make it back, knowing that someday you won't, waiting for that moment."

I look at Delia sitting on our bed, her face swollen from crying, her eyes shut tight. Taking her in my arms again, I try to squeeze the poison out of her. But there is this thing between us now, this difference in how much risk we should take to rid the valley of poachers. I can no longer get as close to her. She is slipping away from me.

"And so, I am leaving," she says, shoving me off again. And I am suddenly more afraid than I was of the storm.

"Look, Boo, we can work this out . . . ," I try again.

"No," she interrupts. "I'm going to the Luangwa River to build a little camp of my own, a place where I can radio track Bouncer and his pride; do something more constructive than sit around waiting for you to kill yourself."

"That's fine," I say firmly. "But it's a long way from here to the Luangwa. Your portable radio won't reach camp from there, and you won't have anyone to drive you out if something goes wrong. What happens if poachers attack your camp? Or if you are bitten by a snake, get malaria, or sleeping sickness? There won't be anyone to treat you or get you out."

"Don't talk to me about risks! It won't be half as risky as flying through treetops at night with people shooting at you, or taking off in a thunderstorm without any instruments."

"Okay, I know, but those are necessary and . . ."

"So are the risks I'll be taking!" she counters. "I have to find a life of my own, so that if something happens to you I'll have some reason for living. Can you understand that?"

Looking at her face, drawn and haggard under the harsh glare from the light over the bed, I can see her aging before my eyes. I cannot remember the last time we laughed, really laughed, together.

"I understand, Boo. Do it if you have to." I have to let her go now, or lose her forever.

The next morning I take off again, to fly a patrol over the northern sector of the park. When I get back to camp, one of our trucks is gone and the kitchen and the office are closed up. Delia has already packed and gone. As I head down the footpath through camp, Marula-Puku seems hollow, empty of spirit, deserted. That night as I get into bed, I discover her note propped against my pillow.

> Dear Mark,
> I love you. Maybe if we survive this, we can start over.
> Love, Delia

The next morning I'm off on a three-hour flight to Lusaka, to have our airplane inspected. Halfway through the journey a strange sickness comes over me: I am feverish; my arms are suddenly as heavy as lead; and my heart is pounding. I'm not fit to fly and should land, but there is no serviceable airstrip much closer than Lusaka. I take Zulu Sierra down to a thousand feet above ground, so that I can make a quick forced landing if I begin to black out.

Dizzy, my vision blurry, I somehow make it to Lusaka, where I rest at a friend's home for more than two weeks. Every time I decide to fly back to camp, I change my mind because I am too ill.

One morning Musakanya shows up with an urgent message: Chikilinti, Simu Chimba, Mpundu Katongo, and some others have headed into the park for poaching. But this time their targets are not elephants and buffalo: they intend killing Delia and me. Not knowing whom in Mpika to trust with this information, Musakanya has hopped the mail bus to Lusaka to deliver his message in person.

The director of National Parks immediately orders trusted scouts from the Southern Province to Mpika and into the park to secure our camp. Musakanya will go back with them. But it will take two days before they can get there. Meanwhile, Delia is on her own at the river all the way across the park. Even after they arrive at Marula-Puku the scouts would not be likely to go to her, and anyway there will not be enough of them to cover camp and the airstrip and protect her. To find Delia the poachers have only to follow her truck's spoor through the bush to her camp. She will be an easy and satisfying target for them.

Still dizzy and weak, I take off in the plane headed for camp. I feel okay until about an hour and a half into the flight, a little more than halfway there. Then my head begins spinning, my vision blurs, and sweat breaks out on my forehead. I force myself to relax, closing my eyes for seconds at a time, and I just keep flying. Over and over I imagine Delia camped alone on the Luangwa, unaware that poachers may be coming to kill her. For the first time I can understand the fear she has lived with, watching me fly off to engage the same poachers over and over.

Nearly three hours after takeoff I fly low over Marula-Puku. Banking over the airstrip I notice a foxhole that the scouts must have dug near my avgas dump. Neither Delia nor her truck are in sight. I firewall the plane's throttle to get full power and head for the Luangwa to find her camp. Fifteen minutes later the river's broad sandbars and pools of hippos flash under the plane. Rolling steeply to my left, I drop between the rows of tall trees that line its bank and fly upstream just above the water. She must be here somewhere . . . Damn her for going off on her own at a time like this!

I spot her khaki tent with its brown flysheet, tucked under a grove of ebony trees on a tall bank above the river. But there is no truck in sight, and no Delia.

"Brown Hyena, Brown Hyena, this is Sand Panther," I call over the radio, using our call signs from the Kalahari. "Do you read me, Boo?" Flying back and forth over her camp, I call again and again. But she doesn't answer. I hope she has her radio.

I try the radio again: "Delia, if you can hear me, listen carefully. Poachers have come into the park to kill us. Stay at your camp and I will be there later tonight." With little daylight left, my heart heavy with worry, I follow her truck's spoor toward Marula-Puku, hoping to spot her somewhere along the way.

A minute later my radio crackles to life. "Sand Panther, this is Brown Hyena. Mark, is that you? I've been away from camp doing some survey work. Over."

"Roger, love. Now listen . . ." And I tell her of the poachers and my plan to drive out tonight to escort her back to Marula-Puku.

"Negative. There's more work to do here. I have my pistol, five rounds of ammunition and I have a game guard with me. I don't need you to come rescue me. See you at Marula-Puku on the weekend." Nothing I say changes her mind, so I head for Marula-

Puku. Later we will learn from Mpundu Katongo and Bernard Mutondo themselves that they, Chikilinti, and Simu Chimba left Mwamfushi on foot one night, each armed with an AK-47. Three days later, at four in the afternoon, they crossed the truck track two miles west of Marula-Puku. They walked a little farther, then camped for the night on a small stream between two hills.

At sunrise the next morning they made their way through the bush and up the back side of the hill above camp. With Chikilinti in the lead, they crawled to the edge of the bank and peered down on the stone cottages. The cottage on their right, where the trucks were parked, was closest to them.

They watched me walk along a footpath and into a building near a campfire. For the next fifteen minutes they could hear me talking on the radio. Then they saw me come out and begin walking toward the building where the trucks were parked.

The four men crawled along the steep bank until they came to a stream cut, then scrambled down to the cover of the tall grass below. From there they watched as I walked into the workshop with three of our men. They crawled forward through the grass until they were within seventy-five yards of the trucks. There they waited, lying side by side in prone shooting positions, fingers on their triggers.

When I stepped from the building, they quickly sighted along their weapons at me. "Wait!" whispered Chikilinti. A game scout with a rifle slung over his shoulder was talking to me, then another stepped into view. Raising his head and leaning slightly to the right so he could see past the trucks, Chikilinti saw five more game scouts sitting around a campfire. Their berets and AKs told him they were special paramilitary scouts, not his game-guard friends from Mpika and Kanona. Their presence was totally unexpected.

I climbed into the Land Cruiser and drove away from the cottages toward the hill where the plane was kept. The poachers backed away from the edge of the camp and quickly retreated up the bank. They could not risk attacking with the scouts present. Under cover of the tall grass and bushes, they ran for the airstrip.

The poachers crossed a stream, then climbed through mopane scrub to the crest of the next ridge. From the peak of a termite mound fifty yards off the runway, Chikilinti could see me and our two guards fueling the plane. A rifle was leaning against the tail of the plane and I had a pistol in a holster on my hip. The Bembas were unarmed. Chikilinti and his men crawled through the tall grass to a point directly across the airstrip from the plane and only sixty yards from it. Flattening himself to the ground, he pointed at something. Simu Chimba wriggled up beside his leader and looked along his finger. Just twenty yards behind the plane, a head covered with branches peered out of a hole in the ground near a stack of avgas drums. The stubby muzzle of an AK was visible over the small mound of dirt. Looking carefully, they discovered several other scouts dug in around the strip and three more walking out of the guard house toward me.

Withdrawing through the scrub brush, the poachers turned and hiked as fast as they could back to Mwamfushi, pausing only briefly at sandy river crossings or dusty game trails to brush out their tracks with small mopane branches, in case they were being followed. Later, in hiding near the village, they sipped beer from a good mash and made plans to attack the bwana's camp after the game guards had gone back to their posts.

18
Nyama Zamara

DELIA

> I make all my rounds
> without leaving a trace
> and sit by the water
> the breeze in my face;
> The future is distant
> yet triumph is near
> I notice the sounds
> only spirits can hear.
>
> — SETH RICHARDSON

o o o

I MAKE MY OWN CAMP on the Luangwa River; the Bembas call it Delia Camp. It is tucked in a *Combretum* thicket under the dark, embracing arms of an ebony tree, overlooking a broad bend in the river. So many massive trees line the torn riverbank — eroded during the rains by raging floods — that my small grass hut and tent are completely hidden. A sprawling white beach fifty yards wide reaches toward the water. Crooked snags and the remains of entire trees — swept away by the torrential rains — lie like fallen monuments in the now-forgiving current, which laps gently at hippo ears.

This land is wild land. Deep ravines, jagged tributaries, and lush oxbows are lost in green forests draped with intertwining vines. Powerful roots of the strangling figs smother towering trees. This is not the open savanna I have seen; it is what I thought Africa was all along.

It has been no easy task to reach this spot. Days of breaking trail and mending tires have brought me once again to a rugged paradise. I have come to see if what we are fighting for still exists; have we become so immersed in the battle that we do not realize it is already lost? I have come to see if Africa is still here.

It is. Hundreds of hippos — the river lords — laze, yawn, and sleep just beyond my beach. Puku, warthogs, and impalas graze on the far shore. On one walk I see zebras, kudu, waterbucks, buffalo, and eland. One morning a large male lion walks into my camp, and the next day a leopard saunters along the same path. I can hear baboons somewhere behind me, scampering and foraging their way through the forest. And a Goliath heron glides by on slow, silver wings.

With the help of the Bembas I build a small grass hut on the riverbank, overlooking the beach. It is a frame of deadwood logs tied together with strips of bark and covered with a slanting thatch roof. One end is enclosed with neat grass walls, where I hang baskets, pots, and pans. The tin trunks of supplies are stacked against the walls, and grass mats cover the ground. The other end is open like a veranda with a sweeping view of the river. Here I have my table and chair for working and eating. I always return from a hike with a bird feather, porcupine quill, or snail shell, which I stick into the grass walls until my little hut looks more like the nest of a bower bird than a house.

Hippo paths go around both sides of my tent. At first this seems like a good idea — to be close to where the hippos walk. But I have to stretch the guy ropes across their paths, and when I lie down to sleep on my mat, I worry about what would happen if a hippo tripped over the rope. Glad that Mark is not here to laugh at me, I wriggle out of my sleeping bag and remove the ropes. I would rather risk the collapse of the tent from an unlikely windstorm than a ton of mad hippo stumbling over the cord.

My friend the moon is here, casting his warm illumination on the river bends. With ease I watch the hippos at night as they leave the river and wander along the beaches toward the grasslands and lagoons. Although some males have territories along the river and forage there every night, none of the hippos — including the territorial males — use the same paths every night. In fact, they do not sleep in the same part of the river by day.

Before sunrise one morning, when the river and banks are shrouded in fog, I watch an enormous female hippo waddle across the beach with her tiny calf. When their young are born, females separate from the group and tend their infants in secluded reed beds for several months. This female appears to be returning to her group with her youngster, which bobbles around the beach like a rubber toy. But when they reach the water's edge, the baby halts. The mother looks at the calf and I can almost hear her saying, "There's something I forgot to tell you. We don't live on the beach; we live in the water."

Curious to see how a baby hippo gets into the river for the first time, I creep through the grass to get a closer look. The mother eases into the current, until her legs are half submerged, then looks back at her young. He follows, but is soon completely under water except for his head. The mother lies in the shallow water with the baby bobbing next to her. Later, when the sun is hotter, the mother pulls herself into the deeper current, and her offspring floats with his head resting upon her titanic nose.

While walking one cool July morning only a few days after setting up my camp, I discover a lagoon that stretches into a maze of clear waterways, so thick with life that the surface seems to breathe. Almost every inch of the still water is covered with bright green lilies and "Nile lettuce." The lagoon is only a half-mile from camp, but so well hidden in the forest that I do not see it until I am twenty yards away. If I had been more clever, though, I would have known it was here by watching the ducks and geese fly over my camp every morning in this direction. After they have roosted all night in the cold air, their bellies are nearly empty and they waste no time flying directly to the pools, where a buffet breakfast floats on gentle waves. In the evening — their bellies full — they fly back to their roosts on a more leisurely course, following the river bends, as though taking the scenic route home.

Stepping silently through the undergrowth, I walk almost every day to the lagoon and am never disappointed. Crocodiles slither from the banks into the emerald shallows, where a few hippos lift lily-covered snouts to peer at me. Hundreds of ducks, geese, storks, plovers, herons, eagles, coots, owls, lily-trotters, and hornbills wade, waddle, feed, and call in a natural aviary. Nile monitor lizards — more than five feet long — splash into the water and take refuge under the weeds. Puku, waterbuck, impalas, and buffalo nibble succulent morsels in the sunken glades interlacing the lagoon. Once I come so close to a female bushbuck and her fawn that I can see the sunset in her eyes.

This is not a swamp, oozing with decomposing matter; the water is as clear as a glacial lake — except, of course, where a thousand webbed feet have stirred up the bottom. The morning grows tall and hot, and the surrounding mopane forests are dry. I want to sidle up to the lagoon's edge, put my toes into the coolness, and slide into the pure water. I imagine that my hair would drift among the lilies and that all my troubles would dissolve. But the jagged ridge of a crocodile's tail glides slowly back and forth, just below the lily pads, reminding me that this is not my lagoon. Still, I cannot resist tiptoeing a bit closer and leaning over to touch the water. Stiff-necked and poised to spring, I must look like an impala approaching cautiously to drink. Yet I am held back from the peace and comfort of the lagoon by a strong fear; it is not my reflection that I see in the water, but the reflection of Africa beyond me.

The lagoon has no name. In fact, I cannot find it on my twenty-year-old map, so perhaps it is only recently born of the river. I ask Kasokola and Mwamba, who are with me, to help me think of a beautiful African name for it. After a few days they suggest "Nyama Zamara." Liking the sound of the phrase, I ask where it comes from and what it means.

"It is in the language of the Senga tribe from Malawi," Kasokola tells me. "It means, 'The game is finished.' "

I am taken aback. "But that is a horrible name for such a beautiful lagoon."

"It does not mean that the game is finished in this place," Kasokola explains. "But once a poacher was caught here, and on his clothes he had written the words, 'Nyama Zamara,' so that is what the lagoon must be called."

Reluctantly I accept the name, but add, "Please, you must help me make sure that the wildlife is never finished here. We must make this pledge." And they agree.

For despite the abundance of wildlife, poaching is a serious problem here, as elsewhere in the park. We find twisted wire snares on the game trails just outside camp, and a tree blind poised over the lagoon, where a sniper must have taken his pick of the gentle creatures who came to drink. One of the reasons I have come is to establish a game guard camp, the first inside the national park.

Of course, we could have sent the game guards on their own, or with our staff, to set up a post here. The main reason I am here is that at least for now, I can no longer live with Mark in base camp, which has become the command center for antipoaching operations. Hordes of green-clad game guards file through on their way to forced patrols. Mark flies them to remote airstrips, packs them off toward poachers' camps, drops supplies to them from the air — yet only rarely do they return with poachers. The last time he supplied them from the air, risking his life as he flew right over the treetops, it turned out that all they wanted were cigarettes. Over and over he puts himself in extreme danger for these men, but it is obvious that they are not going to do their jobs. I beg him to try something different — for example, hiring carriers to supply the game guards, or even training our own scouts. He believes strongly, though, that it is the game guards who must ultimately protect the park, so we must do whatever it takes to get them working.

When Mark flies dangerous night missions — in and out of storms — searching for poachers' campfires, I am the one responsible for organizing an air search if he does not return. Still, when I ask him when he expects to be back, already racked with stress he snaps that he has no idea. To make things worse, the radio is very unreliable. Sometimes I can hear him relay his position, then suddenly, as he is talking, the radio goes dead. Pacing back and forth under the marula trees in the middle of the night, I don't know whether he has crashed or not, until I hear the drone of the plane sometimes hours later. To get away from this madness, I stay busy with the village work and the education programs. But it is not enough, and finally I decide to set up my own camp to protect this one corner of the wilderness.

I know that our strong love is somewhere just below the surface, but I also know that love cannot survive unless it can grow. I can't afford to let myself feel too much for Mark, for tomorrow he may be dead. We have to find a way to face this struggle together, or we may come out apart on the other side — if we come out at all.

The game guards' field camp is almost a mile north of my camp. Since they will not have to walk all the way from Mano into the park, I hope they will patrol more often. They are supposed to use their tent and food stores as a base from which to conduct four- or five-day antipoaching patrols throughout the area. But whenever I drop by, I find them lounging around the camp. I give them pep talks and cigarettes; there is no improvement.

If I wake up early enough, at about five o'clock, and look out of the tent window, I can see the hippos strolling across the beach on their way back to the river. If I sleep too late, all I see are ears and noses in the current. It has become a game with me. As soon as I open my eyes, I scramble out of my sleeping bag, unzip the tent door, and count the hippos.

One morning I hear thumping on the beach as an animal runs across the sand. Hurriedly I step out of the door and find myself face to face with a lioness. She has loped up the hippo path and stands only five yards from me. Sleek and golden, she twitches her tail ever so slightly and looks me over. Then she looks back at the beach, where two other lionesses and four cubs saunter toward us. Without even a glance at me, the others walk up the hippo path past my tent and into the mopane forest.

Pulling on my clothes, I rush to the tent where Mwamba, Kasokola, and a game guard are sleeping. Calling softly, I ask them to come quickly to follow the lions.

Walking along the hippo path, we soon catch up and trail the lions, staying about one hundred fifty yards behind. The females move purposefully, as though they know exactly where they are going; the cubs roll and tumble behind their mothers, playing chase through the grass. Before we reach the edge of the trees, the females stop and look ahead. Standing in a small grassy clearing is a large male lion with a full golden mane. As I look through my binoculars I see that he wears a radio collar; it is Bouncer. It has been more than a year since we've seen him. Because of the poaching war, we haven't been able to study the lions as we had wanted.

The females rush to Bouncer and greet him with long, sensuous rubs. Lifting his tail, he scent-marks a bush. Then they settle into a deep *Combretum* thicket to spend the day in the shade. We humans return to my little camp for breakfast. The vegetation is too thick, the terrain too rough to follow the pride in the truck, and I do not feel safe following them at night on foot. But each morning and evening I search for them and make observations on their prey choice and habitat movements.

Now that it is the dry season, they have moved away from the plains and spend most of their time in the woodlands near the Luangwa. One morning I watch the females hunt a warthog; another morning they stalk a puku. But mostly they kill buffalo, one of which is large enough to feed the whole pride for several days.

The days are cool now, and one day the lionesses and their cubs spend the entire morning sunning themselves on the beach. Soon a female puku and her fawn walk across the sand toward the river. She apparently wants to drink, but seeing the lions she stops about fifty yards away. For a few moments she looks back and forth from the water to the lions, and ultimately lies down with her fawn right where she is. Watching from my camp, I wish so much that I could stroll across the sand and join the ladies on the beach.

The breeze blows strongly against my face, and as I stand alone on the riverbank, I feel as though I am being interviewed by the wind. How can we hold on to the old, wild Africa? When the elephants are safe, can we go back to our lives of lion watching?

I long to return to studying the animals, sinking into Nature and learning her ways. There are always more wonders to uncover. Discover them quickly, before they go — is that where we are? In the Kalahari, Mark and I discovered the second largest wildebeest migration in Africa. No one else saw it except the sun; a few years later it was gone. Will anything save the elephants? Will the rain bring back the desert? Will the desert bring back the wildebeest, or have they all marched on to a world — a time — that was and will not be again? Somewhere, is there a dusty plain where wildebeest still dance?

The wind rises and slaps my hair, but I do not have the answers.

The Africa I speak of is still here, but it is in little pockets — in small corners of the continent — hiding.

o o o

Driving back and forth between Mpika and the Luangwa, I divide my time in the dry season of 1990 between our work in the villages — that we now call the Community Service and Conservation Education Programs — and my studies at the river. We have expanded our programs to ten target villages that are near the park and have many poachers. These projects have grown so much that we have taken on a full-time employee, Max Saili, and two volunteers from Texas, Tom and Wanda Canon. I still believe that the only way to save the elephants in the long term is to convince the people in the area that they are worth saving.

On one of my trips back to base camp to resupply, Mark and I walk along the Lubonga, trading war stories about the game guards. He has not had time to visit my camp, and the only chance we have to see each other is when I come to Marula-Puku for supplies. There is still a strain between us and we operate more or less in our own realms, our personal lives on hold.

Mark lags behind me, which is so unusual that I turn around to see if he is all right. He walks slowly, his head down, the .375 rifle slung loosely over his right shoulder.

"Are you okay?"

"I feel a little weird." Beads of sweat glisten on his forehead. "Need some food. Gotta get to camp, some food." He starts to sway and stumble.

Rushing to him, I grab the rifle. "First, you've got to sit down." I push him gently to the ground. He sits hard and then falls backward. His right arm flings out straight, his head lurches back, and he slumps into unconsciousness.

"Oh God! Mark, is it your heart? Tell me what's wrong!" His eyes are open, staring blankly. I grab his clammy wrist and feel for his pulse. It is strong. His breathing seems normal but his lips are blue. Mark has had mild trouble with his blood sugar for years, but can usually control it by eating properly and avoiding too much caffeine. Lately, however, he has been drinking cup after cup of strong black coffee with sugar to stay awake for his hours of night flying. Since his sickness in Lusaka he has been lethargic, barely able to keep going. I have begged him to see a doctor, but he has refused, saying that if anything serious were found, he might be grounded, and that his flying is our only weapon against the poachers.

Now he lies totally unconscious, legs and arms splayed across the ground. I try to stay calm, to think. I have to get some glucose into him. I run the five hundred yards to camp, screaming for Simbeye to get the truck. In the kitchen boma I mix powdered milk, water, and honey into a canteen. My hands shake, and white powder and goo spill all over. What seems like hours later, all the guys are in the back of the truck and we race out to where Mark lies.

I stare deep into his vacant eyes, which seem pale and lifeless. If I try to give him the honey mixture now, he will choke. Since it is Sunday, there is no radio schedule until tomorrow morning — no way to call for help, and a six- or seven-hour drive to the clinic. It has been twenty minutes since he collapsed.

Slowly he blinks. He moves his head. I put my face directly over his.

"Mark, can you see me?" He twists his head around, his eyes full of fear.

"What happened? Where am I?"

"It's okay. You fainted. I mixed some milk and honey for you. Can you drink it?"

"I think so." But before I can get the canteen to his lips, he passes out again, staring straight ahead with empty eyes for another ten minutes.

Again he tries to look around him. "Boo, where am I? Did I crash the plane?"

"No, Mark," I say softly. "It's okay. You're right outside camp. Remember, we were walking."

"Oh, I see." He faints again. For fifteen more minutes he slides into and out of consciousness.

Kasokola and I lift him by the shoulders, until he is slumped against me. When he wakes again, I murmur, "Here, drink this."

He sips the milk laced with swirls of honey. Leaning heavily against me, he is able to get five or six swallows down. He rests with eyes closed.

"I feel better," he whispers. I push the canteen to his lips and he drinks again. "Where am I? Is anybody else hurt?"

I explain again, holding him tightly in my arms. "It's all right, everything is all right."

After ten minutes he is able to sit up by himself, and is already joking with the guys, who stare from the back of the truck with deep concern on their faces, like so many masks. "I just had too many beers," he cracks, "what's all the fuss about?" But they do not laugh. To see a strong man in such a state is not funny.

I call softly to Kasokola and Simbeye, and we lift Mark into the back of the truck and lay him down. I drive him to the bedroom cottage, and they help me get him into bed. Kasokola asks if he can bring more milk and honey, and I say, "Yes, please."

When they are gone, I whisper, "Mark we can't go on like this. We have to do something different. We have to get some help."

"You're right, Boo, we will. I promise."

But the next morning — only fifteen hours after passing out, and in spite of everything I say — Mark flies another antipoaching flight. Feeling that I can no longer reach him, I return to my river camp.

19

Close Encounters

MARK

> The elephant moves slowly to protect its vast brain,
> With which it hears subsonic sound,
> And in which it carries the topology,
> The resonances and reverberations,
> Of a continent.
>
> — HEATHCOTE WILLIAMS

o o o

I HAVE JUST FALLEN ASLEEP when I am awakened by the
sounds of harsh breathing, heavy footsteps, and grass being ripped
from the ground somewhere near my head. Wafting through the
window is a sweaty, bovine odor mixed with the sweet pungence
of marula fruits. Sliding slowly out of bed, I press my face to the
flyscreen. Six Cape buffalo loom in the darkness, one of them an
arm's length away, its stomach churning like an old-fashioned
Maytag. Moving along the cottage, the old bull rakes his horns
against the stone wall, making a clacking sound. I am lulled to
sleep knowing that in some small way we have been successful.

o o o

When we first stood on the high bluff looking down on Marula-
Puku, we didn't know it had once been a poachers' camp. But we
soon found abandoned meat racks and ashes beneath the marula
trees. Poaching had conditioned the animals along the Lubonga to
fear humans. After we arrived and began defending them, it took
only a few months for the animals to grow accustomed to their new
sanctuary. Soon puku, impalas, wildebeest, buffalo, and waterbuck
grazed across the river from us, then around camp, and finally
among our cottages at night after we had gone to bed. Mornings

we would hurry outside to see how the shrubbery had been rearranged during the night.

For some reason the warthogs seem more shy than the buffalo and other animals. With faces like uprooted tree stumps — all knobs and nodules — they have a right to be shy. But during the dry season of 1990 a boar, a sow, and three piglets begin feeding across the river from camp, usually in the late afternoon. On their front knees, they root for tubers along the far bank, occasionally splashing through the rocky drift upstream from camp, all in a line, their tails stiff as pokers and straight up like lightning rods. Whenever they reach the edge of camp, they stop to watch us for a while, then trot off in the opposite direction, still afraid of us after all this time.

One afternoon in late August, Survivor comes into camp again on his way to the mountains. There are no marula fruits now, but he feeds on the new leaves and seeds of the *Combretum* trees. The six buffalo still graze along the bedroom cottage each night until morning, but otherwise Survivor has camp all to himself.

Right on schedule, he strolls into camp along the track. Jogging close at his heels, as if they were his miniature cousins, are the members of the warthog family. The gray, wrinkled pigs trot in the footsteps of the gray, wrinkled elephant. As Survivor stops to feed on some small shrubs, the warthogs fan out to root tubers, shoots, and bulbs. They never venture more than twenty feet or so from their towering companion, and when he finally ambles out of camp, the warthogs jog in a line behind him, down the track and out of sight. If only Delia could have been here to see this procession.

Soon Survivor is coming every day with his entourage, and as I walk along the footpaths between the kitchen boma and the office cottage, the pigs pay no attention to me and keep on rooting.

o o o

One evening Luke Daka, permanent secretary to the minister of tourism, and Akim Mwenya, deputy director of national parks, are sitting with me on the riverbank at camp. Over the years Delia and I have met with numerous officials in Lusaka to describe the poaching and corruption, and have invited them to visit the project. Daka and Mwenya are the first to come.

"Look! An elephant!" He points across the river as Survivor strolls to the water's edge. Jumping up from his chair, he exclaims, "I can't believe it. I am Zambian, but this is the first time I've ever seen an elephant in the wilderness."

"That's Survivor. If you can imagine it, there used to be seventeen thousand elephants in North Luangwa. Now, because of poaching, maybe a thousand or two are left."

"That's awful," he says. "Of course, you have told me at our meetings in Lusaka, but I didn't imagine it was so bad."

"It's bad all right. When I next see you in Lusaka, Survivor may be dead — unless you can help us." I tell him again about the apathy and corruption among the game scouts and officials in Mpika. By the time I finish, Daka's face is sagging. He promises to do what he can.

The next morning, after Daka and Mwenya leave, Survivor is back in camp. The poaching is still bad, and I should be flying. Delia is right, though — I need a day off. So I gather up my daypack, field notes, and binoculars, intending to follow my favorite elephant, record his behavior, and see where he will lead me. I'll have to do this surreptitiously, because Survivor still does not like people to approach him; he prefers to make the advances.

He leaves camp by way of our outhouse, where I fall in behind him. I stay far enough back that I can keep his tall rump in sight without his seeing me, and stay downwind so that he won't be able to smell me. He crosses the stream cut next to camp, climbs a steep bank, and wanders through an expanse of *Combretum fragrans*, casually wrapping his trunk around an eight-foot bush and decapitating it as he passes. He doubles back to our track, crosses it, and descends the old false bank of the Lubonga, where the river used to run, there to stand in a deep glade under a huge marula tree. I close to within thirty yards of him, sit with my back against another tree, and we doze together until midafternoon.

At about three-thirty he leads me along the river, past the airstrip ridge and onto a floodplain near Khaya Stream. The grass here is more than eight feet tall. Since I can no longer see him, I follow his rustling sounds.

Other elephant paths intersect this one now, so I quicken my pace, afraid that he has taken one of the trails to the side and is about to lose me. I pause to kneel at a pile of dung and an elephant's track at the junction of two paths. The grass in the track is flattened to the ground, but a few stems are beginning to rise as I watch. I poke my finger into one of the balls of dung; it is very warm. He has just passed. I stand and hurry on.

Suddenly the air reverberates with a deep rumble, like thunder far away. There is Survivor, less than ten yards ahead, curling his trunk high above the grasses like a periscope. Other deep rumbles sound. And now I hear the grass rustling from several directions. Other elephants are coming toward us. Afraid to retreat, I sidestep off the trail into a shallow mud hole and squat down. The massive crown of a bull elephant, ears flapping, feet swishing in the grass, cruises by so close I can see his eyelashes. He is like a giant combine harvester in a field of tall wheat. Reaching Survivor, he lowers his head and the two bulls briefly push at each other. Several other bulls arrive, milling about Survivor for a minute, their trunks touching the streams of temporal gland secretions that flow from the sides of his head, behind and below his eyes. The elephants remind me of humans, shedding tears of welcome. Then Survivor leads them off toward Khaya Stream two hundred yards ahead. I follow.

Nearer the stream the grass mostly gives way to tall *Khaya nyasica* and *Trichelia emetica* trees, which stand on a high bank above the streambed. The elephants follow the trail through a deep cut in the bank and disappear. Creeping closer, I hear the sound of splashing water. Anxious to see what is going on, I crawl into a clump of grass on the bank and ease forward, my chin on the ground, until I can see over the edge.

Below, the elephants have assembled around a pool at the edge of the sandy streambed. A ledge of rock running across the stream provides a shallow basin that collects water trickling from a natural spring halfway up the bank. They stand shoulder to shoulder around the basin, drawing up water, raising their heads and curling their trunks into their mouths to drink. Their thirst slaked, they squirt water over their own backs, and onto one another. It is an elephants' spa.

Finished with their bath, they walk five yards to a dust wallow near an enormous ebony tree that grows horizontally about six feet above the ground. With the tips of their trunks they gather a quart of the gray powder at a time, flinging it over their backs and between their legs until a great cloud rises against the red sun of late afternoon. Then, one by one, they file past the "Scratching Tree," each leaning against it at a spot rubbed shiny smooth, heaving his bulk up and down against its bark, his eyelids heavy.

For the next half-hour, they stand around in the spa, heads hanging, trunks resting on the ground. My neck tires of holding up my head, so I rest my chin on my hand. But in moving I dislodge a pebble that rolls down the bank into the streambed near Survivor. Jerking his head up, eyes wide, the bull elephant pivots toward me. His trunk is up and air blasts from his mouth. Aroused, the others spin around, prepared to flee. Lying motionless, my face in full view, I feel like a Peeping Tom. Survivor walks slowly toward me, his trunk snaked out. Before I can react, the tip of his trunk is snuffling through the grass, inches away from my right foot. The tall bank stops his advance, but he keeps reaching for me, his trunk fully extended, patting around in the grass, as though looking for my foot. We stare into each other's eyes for long seconds. I think he knew I was here all along — just another guy in the locker room.

Curiously, Survivor takes a bite out of the bank, withdraws his trunk, turns and walks back to the others, dropping the grass and soil from his mouth along the way. Alert, but apparently no longer afraid, the elephants file into the long grass beyond the stream, Survivor in the lead.

The shadows are long, so I follow an elephant trail along the Lubonga on my way back to camp. Not far from the elephants' spa, the path leads to the edge of the Lubonga, where the river cuts into a low hill on the opposite bank. Directly across from me a bull buffalo is grazing the grass on a shelf fifteen feet above the water. Immediately beyond him the hill rises steeply away from the river. I am so close to him, no more than thirty yards, that I am amazed he has not seen me. Raising my field glasses to watch him, I immediately realize why he has not: his left eye is bluish-white and stares sightlessly into space. His hide is covered with old scars, his ears have been shredded by thorns, his tail has been stumped — probably by lions or hyenas — and over the years the African bush has worn the tips of his horns to polished black stubs. He has been well used — almost used up.

Still unaware of me, he continues to graze, moving on a diagonal toward the sheer edge of the tall bank — and in the direction of his blind eye. Step by step he grazes closer. Finally he puts his left front hoof within six inches of the edge and a slab of the bank crumbles under him. The sixteen-hundred-pound buffalo pitches over the side, feet flailing, into the shallow river below.

Heaving his bulk onto his hooves with surprising agility, puffing like a steam engine, the old buff swings around and charges the bank, trying to get back up on his ledge. Scrambling halfway up, he loses his footing and rolls into the river again, spray flying. He tries again, with the same result. And again. Each time becoming more frantic — and more pitiful.

But before I can feel very sorry for him, the buff spins to face me. Standing with his front legs splayed, he swings his head from right to left, taking in his options with his one good eye: obovatum thickets to his right, another bank to his left, and one behind him. I am his path of least resistance; I might as well have cornered him in a back alley. Somehow the water between us has given me a false sense of security. But here the river is inches deep, as he is about to demonstrate.

Bellowing and snorting, he charges through the river, fountains of water erupting from his feet and crashing around his body. About fifteen feet behind me is a dead tree with a limb at eight feet above the ground. I whirl and run for it, the bull's hooves thumping hollow on the ground behind me. Praying that it will hold me, I jump for the limb, hauling myself up and out of reach as the buffalo charges past and out of sight beyond a thicket. For several minutes I perch there, catching my breath. When I am sure he's gone I jump down, smiling to myself, and head for camp — glad that Delia did not see me running for the tree. She probably would have shot me in the back.

20

The Last Season

MARK

... an old dream
Of something better coming soon
for each Survivor.

— L. E. SISSMAN

o o o

I AM WALKING PAST the big marula tree near the center of camp
when Mumanga's "pssst" stops me in my tracks. Looking up, I
see a bull elephant watching me from forty yards away. I back up
to the wall of the office, and after hesitating for several minutes,
the elephant walks slowly to the kitchen. He stands ten feet from
the fire where Mumanga is cooking, searching with his trunk
through the dried leaves for fruits, curling one after another up
into his mouth, his ears flapping back and forth. I walk closer, until
I can see the hole in his left ear. It is Survivor.

At first Mumanga stays rooted to his spot beside the stove. But
as Survivor moves even closer, he slowly retreats, backing up until
his heels strike the step to the kitchen door and he sits down with
a plop. Our wood stove and cooking area is covered with a large
thatched roof, which gives some shelter against the sun and rain
— but not against elephants. Apparently deciding he needs some
roughage to go with his fruit, Survivor snakes his trunk to the
thatch, pulls out a great plug, and stuffs it into his mouth. The
poles supporting the heavy roof stagger, crack, and groan until I
think the entire structure will collapse. And so does Mumanga,
who darts into the tile-roofed kitchen cottage. Fortunately Survivor does not care for the taste of the coarse dry thatch; he spits it
out and picks up another fruit, feeding around the kitchen and
workshop until after dark. Then he shambles away from camp

through the long grass, like a massive gray boulder rolling through the moonlight, his great round feet leaving tracks across my life.

When I get up for breakfast, one side of the thatched roof of the open kitchen boma is sitting on the ground; the other is perched on top of the smashed stove. Weakened by Survivor's assault, the poles supporting the roof have collapsed during the night — with a little help from termites. Delia will have a surprise when she gets back from her camp on the river.

Later that morning I fly to Mpika to talk with Warden Mulenga about the Mano scouts, who still seldom go on patrol — except when they want some meat. Shuffling her papers and grinning shyly, Mulenga's secretary tells me he is out for the day. Her reaction tells me he is "out" drinking at ten o'clock in the morning. I drive to the British Aid Compound, pick up our mail, and am about to leave for the airstrip when a man named Banda Njouhou, one of the warden's senior staff, catches up with me and leads me to a quiet corner of the compound, where we sit under a tree.

Looking about nervously, he begins. "Mark, the warden is not only warning poachers of patrols, he is poaching himself." On weekends Mulenga is taking trucks donated to National Parks by USAID (United States Agency for International Development) and driving to the game management areas, where he orders scouts to shoot buffalo and other animals. Then he hauls the bush meat back to Mpika and he, his wife, and two scouts sell it from his home, at the market, at two bars, even at the banks.

"The warden uses the Mano Unit's ammunition and gasoline rations for poaching," Banda goes on, "and he sells their mealie-meal rations. He is also trading in animal skins." Patrick Muchu, the Fulaza unit leader, and some of his scouts are shooting cheetahs, leopards, lions, zebras, and other animals, whose skins Mulenga sells on the black market through a senior parks official in Lusaka. He and the official transfer scouts who are Mulenga's cousins to areas around the North Park, to help him poach. And they send away good scouts like Gaston Phiri.

"What can we do about this?" I ask Banda.

"We could catch him at a roadblock some night when he is returning to Mpika with his truck full of meat."

"Yes, and what scouts or police can we trust to arrest him? And if they arrest him, what do you think the magistrate will do? He poaches too, you know."

He holds up both hands. "Yes, I know these problems very well. I don't know what you can do. I can do nothing; these are my bosses."

As we walk back to his truck, I thank him for coming to me with his information.

Almost as soon as he drives away, another truck pulls up and stops near me. Two men get out and introduce themselves as officers from the Anticorruption Commission. One of the men, small, with a round face, frowns. "I am afraid someone has charged you with buying black-market military weapons." The other man, slightly taller, with a thin, haggard face, looks silently at me.

"What? Who says so?"

"Do you know Bwalya Muchisa?" asks the round-faced man.

"Bwalya accused me of this? Where did you see him?" He went to visit his family almost two months ago and I've been wondering what has happened to him.

"He's in Lusaka. He's been caught with a new black-market AK, and he told a minister you gave him the money to buy it. Look, we know about your good work, but I'm sure you understand that some very important people are dealing in ivory and don't like what you are doing to stop it. This minister may be one of them. He wants you out of here very much."

Sitting at the table in our hut, I write six pages of testimony, saying that Bwalya must have bought the AK for poaching, and when he was caught with it, he came up with the story that I had given him the money to buy it. Since his father is one of Zambia's most infamous and successful commercial poachers, he probably has connections to this minister, connections Bwalya has exploited to try to save himself. Or maybe the minister is out to get me. If so, he must have found out that Bwalya worked for me, learned that he had been caught with an AK, and is trying to use this information to get me thrown in jail or out of the country. Later I would learn from my informants that the whole time Bwalya was working for us, he was keeping an AK-47 and lending it to poachers. As soon as I hand over my finished statement, the officers stand to leave.

"I would advise you to get a lawyer," the round-faced man says. "We'll be in touch. Please notify us if you intend leaving the country." We shake hands and they drive away.

Back at camp, I pack and fly to Lusaka to meet with a lawyer, who tells me that until I am arrested, there is nothing he can do. I brief the American ambassador and the British high commissioner in case I need their help, and then fly to Mpika. There I ask friends to radio us if they hear that the police or military are headed for Marula-Puku. Back at camp again, I refuel the plane and keep it on standby for a quick takeoff.

o o o

During September of 1990 Survivor continues to visit our camp, often bringing with him another elephant we call Cheers. One morning, when Delia has come to Marula-Puku to resupply, we are walking along our footpath to the kitchen for breakfast, the golden sunlight spilling into camp through the marula trees. "Mark, look, across the river." Delia points to a group of five elephants feeding near the staff camp. One is Survivor. We watch him pull up big bunches of long grass, then beat the roots against his foot to dislodge the dirt before stuffing them into this mouth. Sparrow weavers hop on the ground near our feet, chirping loudly, and a family of warthogs kneel on the riverbank upstream, rooting for food. It is one of those cool, still mornings when Africa seems to be standing back from the mirror, admiring a last vestige of her fading beauty.

BWA! A gunshot cracks from across the river and echoes off the high bank behind camp.

Delia spins toward me, half crouching, the color draining from her face.

"Oh God, no! Mark, they're shooting Survivor! Do something!"

BWA! BWA! BWA!

"Get behind a tree!" I shout, pushing her toward the big marula next to our footpath and sprinting for the office. I grab the rifle and shotgun from the corner near the bookcase, and strap on my pistol.

"Kasokola! Simbeye! Get to the truck!" I bellow through the office window. I jerk open the cupboard door, claw ammunition from the shelves, and stuff it into my pockets.

BWA! BWA! BWA! BWA! BWA! BWA! Six more shots thunder from across the river. The hundred-fifty-yard dash to the truck leaves my legs rubbery.

"Do you have your revolver?" I ask Delia as I climb into the Land Cruiser.

"Yes! Just go!" She dumps two more boxes of ammo into my lap.

With Kasokola and Simbeye in the back of the pickup, holding on tightly, we race the quarter-mile to the staff camp, driving across the Lubonga, spray and rocks flying. As we near their camp I lean on the horn and whistle for the guys to come. "Let's go! Let's go!" But they are already running for the truck before it stops.

I hand Mwamba the shotgun and Kasokola the .375, then spill ammo for each gun onto the seat of the truck and scoop it into their hands.

"Where did the shots come from?" I ask. I am afraid I already know, but the echo from the high banks along the river has made it hard for me to be sure.

"That side!" Mwamba points directly at the spot where Delia and I had seen Survivor feeding from our camp. I look at these young men. They are workers, not fighters. Only two of them are armed. The other six are carrying hoes, shovels, pickax handles. "You don't have to come with me," I tell them. "These poachers have AKs. It'll be dangerous." None of them walks away. "I won't think less of any of you for staying here. This isn't your job." Still they stand there.

"Right," I say. "Let's go! Kasokola, Mwamba; help me cover the river while the unarmed men cross!"

No such order prevails. Mwamba and Kasokola immediately

leave me and swarm across the river with the others. I lag behind, training my pistol on the opposite bank. We are sitting ducks as we struggle through the current.

"You guys with the guns, stay up front!" I whisper harshly. We are in the tall grass and brush on the opposite bank, and may run into the poachers at any time. Once at the top of the ridge, we sweep upstream along the river, looking for human footprints going to or from Survivor, or his tracks accompanied by blood spoor — drops of crimson in the dust that may warn us he has been wounded rather than killed. With this sweep I am hoping to keep the poachers from taking cover in the thick *Combretum fragrans* scrub, and to flush them out into the open riverbed. I curse their gall, shooting an elephant within sight of our camp. Then a thought strikes me: maybe they wanted to be seen.

"Nsingo!" I grab the arm of one of our new workers. "Is anyone in camp with Delia?"

"No."

"Then you get back there as fast as you can, in case the poachers attack the camp. Mwamba! You, Simbeye, Chende Ende, and Muchemwu carry on with the sweep. Kasokola and I will go with the others to look for Survivor. If you come across the poachers, don't wait for them to shoot first."

My team scours the riverbank east of where I'd seen Survivor. The tall grass along the river, and the dense brush and hard ground back from it, make the search difficult and dangerous. We will be lucky to get out of this without someone's getting shot, or trampled by a wounded elephant. I soon lose contact with Kasokola and the rest of my group. I hope they will not contact Mwamba and the others and open fire on one another.

After half an hour I still have not found any sign of Survivor or the poachers, and I am growing more and more worried that the shots may have been a decoy and that the poachers are in camp now. Giving up the search, I wade back across the river and run to the truck.

As soon as I pull into camp I see Delia, her revolver strapped to her hip, and Nsingo — but no poachers. Thank God, no poachers. I try to convince her to come with me in the plane to look for

Survivor. But she insists that someone must guard camp, and Nsingo has never seen a revolver before.

"Be careful!" I warn as I jump into the truck.

"Be careful yourself. I'll stand by on the radio."

When I am airborne, Delia gives me a rough bearing from camp to where the shots had been fired. Heading in that direction, I have only flown a few seconds when I see the body of an elephant, its tusks hacked off, lying in a deep sandy stream cut just a little beyond where we had been looking. Vultures are already spiraling down to the carcass, and I cannot tell if it is Survivor. Mwamba and his group are within fifty yards of the dead elephant and walking quickly toward it. I stick my arm out the window and motion them back toward camp. The poachers have fled and there is nothing more they can do here. Better to have them in position to defend Marula-Puku if necessary.

After checking with Delia on the radio, I fly to Mpika to get some of the new scouts who recently arrived. I can only hope that they have not yet been corrupted, but in any case there is nothing else I can do. They are my only hope of capturing these poachers.

Leaving the scouts on the airstrip at Marula-Puku, Kasokola and I take off at four-thirty in the afternoon to look for the poachers. After shooting Survivor they will have headed for Mpika by the most direct route that provides them with cover and water — along the Mwaleshi River. We have flown along the river for only fifteen minutes when we spot their camp near the scarp. I circle some distance away, pretending to be interested in another area. A few minutes later, looking through binoculars, I see some familiar green tents and flysheets. I know immediately who owns them: Chikilinti, Chanda Seven, Mpundu Katongo, and Bernard Mutondo.

I fly back to our airstrip on the Lubonga, land, and then take off again with Brighton Mulomba, the scout leader, to show him the camp and the most direct way to get there. From a distance we can see a column of smoke with vultures circling around it. The poachers have killed another elephant and apparently are feeling secure enough to take the time to dry its meat. I can drive the scouts to a point only a two-hour hike from the poachers, who will be virtually trapped by the valley's steep walls. It will be hard not to catch at least some of them.

Back on the ground I give Mulomba and his troops a pep talk. As soon as I have finished, the group leader announces, "Ah, but you see, my fellow officers and I have decided that we are not due to go on patrol for some time yet."

"But this is an emergency," I explain, "not a regular patrol, and according to the regulations you are bound to take action against these poachers."

"Yes, but anyway we are needing salt and . . ." Resisting a powerful urge to draw my pistol, I turn and walk away.

o o o

Back at camp Delia tells me that she saw Survivor running as the shots were being fired. She believes he may have been only wounded, or may have escaped unharmed. But in the past I have often found the carcasses of other elephants several miles from where they were shot, and Survivor was surrounded by the gunfire. He must have been hit by at least some of the bullets. If they have not killed him outright, he will surely die of his wounds.

We hike across the river to the carcass. Sprawled in the sand of the dry streambed, it is already bloated, rotting, blown with maggots, the juices of its decay fertilizing the soil. It has been so mutilated, first by the poachers and then by scavengers, that at first we cannot make an identification. It is a male about his size, however, and I finally conclude that it must be Survivor. But Delia cannot accept it.

"Mark, we can't see his left ear. We don't *know* it's Survivor." She bites her lip and turns away.

I cannot delude myself that he is still alive. To me he has become another statistic in the war against poachers. All hope of saving this valley and its wildlife seems to have died with him. For our own sakes, for our sanity, we must at last recognize that there is a time to quit, a time to admit that nothing more can be done.

But I have not yet played my last card.

21

Cherry Bombs

MARK

Why not go out on a limb? Isn't that where the fruit is?
— RENEE LOCKS AND JOSEPH MCHUGH

o o o

AT SUNSET THE SAME DAY, Kasokola and I take off and fly low-level north along the Lubonga toward the scarp. By using the mountains to cover our approach, we hope to surprise the poachers who shot Survivor.

I have removed the door from Kasokola's side and turned his seat around so that he can see out easily. Cradled across his lap is a twelve-gauge shotgun tied to his wrist with a bootlace, so that the slipstream will not tear the weapon from his hands. The gun is loaded with cracker shells, each of which will project a cherry-bomb firecracker to a hundred yards, where it will explode with a blinding flash and a very loud bang. Cherry bombs are virtually harmless, but Chikilinti won't know that.

At the scarp I climb the plane into the mountains, dodging peaks and hopping over ridges, staying low to keep the poachers from hearing the plane until the last minute. It is almost dark by the time we reach the headwaters of the Mwaleshi River, a few miles upstream and a thousand feet in elevation above their campsite. I pull back the plane's throttle, lower my left wing, and sideslip into the deep river gorge. Sheer, dark canyon walls loom close on either side of us. I drop flaps to slow Zulu Sierra and bank steeply right and left, hauling the plane through hairpin turns as we descend through the gorge.

By the time we come out of the scarp into the Mwaleshi's narrow valley below, the moon has risen, two days from full and halfway up the eastern sky. Now each of the water holes, rivers, and flooded dambos twinkles with moonlight as we pass; occasionally my heart jumps, but the twinkle is silver — not the flickering, sallow yellow of a poacher's fire.

We have just begun to fly along the foothills of the scarp, approaching one of the Mwaleshi's tributary streams when Kasokola leans out of the open doorway and shouts, "Poachers — fire! There!" He points to a large circle of flames near the river. I bank the plane and push its nose into a shallow dive that will end up over the camp.

Kasokola picks up the spotlight. I warn him not to use it for more than a couple of seconds, just long enough to see the meat racks and confirm that this is the poachers' camp. The light will make the plane an easy target. "Eh, Mukwai — yes, sir." I glance over to see him grinning. He wants Chikilinti as badly as I do. "And don't shine the light inside the plane," I yell in his ear. "It will blind me."

I fly past the camp and turn on the landing light. When I can see the crowns of the trees flashing just under the plane's wheels, I reset the altimeter to read our height above the ground around the poachers. I will use this reading as a minimum safety height to help keep us clear of the woodland while making passes over the camp. Banking steeply, I turn back toward the fires and ease off the throttle, cutting our speed to about 55 mph. As we come in over the camp I cross the controls, using heavy right rudder and left aileron. Foxtrot Zulu Sierra skids sideways through the air, turning around the encampment, the stall warning bawling like a sick goose. The rising heat from the fires below lifts and rocks the plane.

"Now, Kasokola!"

He thumbs the switch on the Black Max and the camp is instantly bathed in the 450,000-candlepower light. Steadying the plane, I lean over him and for a second take my eyes from my flying to look down.

Tents . . . fires . . . and meat racks — covered with huge slabs of meat, so large they can only be from elephants. And a pair of tusks is leaning against a tree near the fire, tusks the size of Survivor's.

A pencil-thin trace of white light flicks past my right wing, followed by a popping sound. Tracer bullets! "They're shooting at us! Fire into the trees and bushes; blast their tents if you can. Don't spare the ammunition." Kasokola puts down the spotlight and grabs the shotgun as I pull the plane up away from the trees and begin turning back again. "Okay, get ready!"

Even with her lights switched off, to the poachers on the ground Zulu Sierra must look like a huge bat flitting about the moonlit sky — a nice fat target. And after our two passes, they must be getting the hang of tracking us with their rifles. I circle to the side of camp away from the moon so we won't present quite such a strong silhouette. Then I drop to the minimum safety height that will keep us clear of the trees, hoping to pass over the camp so quickly they won't have time to get off a shot, at least an accurate one. The dim shadows of trees are skimming by just under the plane's wheels, and then the meat fires are below us. I throttle back, kick right rudder and corkscrew the plane above the poacher's camp. A tracer streaks past, and another, much closer.

"Fire!"

And Kasokola answers: pfsst-pfsst-pfsst-pfsst-pfsst! From its barrel the twelve-gauge issues a trail of red and orange sparks as each cherry bomb arcs into the night.

BOOM-BOOM-BOOM-BOOM! Great thunderflashes of light and sound rock the poachers' camp. Kasokola's face is strobe lit as he cackles with laughter, unable to believe that he has caused all this ruckus. BOOM! The last cherry bomb lands in the campfire. Sparks and fiery traces of burning wood rocket through the camp and into the trees like Roman candles going off. One of Chikilinti's tents starts to burn, set alight by the scattered embers of the campfire.

"Happy Fourth of July, bastards!" I shout. Then, "Reload!" I yell to Kasokola as I haul Zulu Sierra through a tight turn. This time we come in fifty feet higher, spreading our cherry bombs over a broader area, hoping to catch the scattering poachers. I jink and sideslip the plane, avoiding their tracers. Kasokola reloads four more times, shooting up the camp again and again until there is no more return fire.

Thinking the poachers may have split up earlier, I climb to a thousand feet above the ground and see another campfire, about two miles downstream. I push Zulu Sierra into a dive and Kasokola switches on the spotlight. A circle of men with guns is sitting around the fire, near a large empty meat rack. They have not yet killed.

I haul the plane through another tight turn. Back over the second camp, I can see the poachers still sitting close to their fire, so confident that they haven't bothered to take cover. Kasokola puts the first cherry bomb at their feet. KABOOM! At the explosion the poachers throw themselves to the ground, crawling into the bushes. On the next pass we shower them with more cherry bombs and move on down the valley, blasting other poachers in their camps on the Lufwashi, Luangwa, and Mulandashi rivers. We stay out long into the night, until clouds began covering the moon and we have to head home.

After sunrise the next morning, Kasokola and I fly to Chikilinti's campsite of the night before. All that remains are the charred and smoldering shreds of his tent and a pile of squabbling vultures.

22

Scouts on the Prowl

DELIA

You never feel it
till it's over —
the relief
at having survived
and the new sun
rising calmly as ever
before your eyes. It's
morning.

— PAULA GUNN ALLEN

o o o

WE SEE NO POACHED ELEPHANTS in North Luangwa for three months. For four months. For five months. By February 1991 Long Ear's and Misty's group of elephants and other family units stroll onto the floodplains, feeding in the open on the blond grasses. Occasionally families come together, forming aggregations of up to two hundred sixty elephants.

One evening Long Ear's group ventures down to the river's edge before sunset. The calf and the youngster — now five years old — gallop through the Mwaleshi, sending sprays of water into the air. Their mothers bathe nearby; but then, caught in some new spirit, they too run in circles through the shallow river. Chasing one another, adults and young alike frolic and romp — something they have not done in a long time. A natural wildness slips back into the valley.

Still receiving coded messages from Musakanya, Mark flies night and day. He greets every poaching band that enters the park with his special cherries, even before they can set up their camps. Musakanya sends word that the poachers are having trouble hiring bearers, because they are afraid of the plane and its explosive bombs. The carriers start trickling into our small office in Mpika, asking us for work. We hire them with money we have raised in the States.* It is as simple as this: Mark chases them out of the park with the plane; I greet them on the top of the scarp with a job.

o o o

Bumping along the bush track one afternoon, I drive across the valley toward Marula-Puku from my river camp. The track takes me through the dusty plains that are spotted with buffalo and through the forest where kudu hide. Earlier Mark radioed from the airplane, asking me to come by camp to talk about something important. When I reach the Lubonga, I stop in the tall grass and walk to the river's edge, where the water tumbles over a small rock shelf and creates a natural whirlpool. I bathe in the sparkling current and change into a fresh blouse and jeans. Sitting on the rocks, I brush my hair in the sun. I drive the truck across the river and down the track into camp, where I park under the marula trees.

As always, Kasokola and Mwamba rush out from the workshop to shake my hand warmly, and Mumanga, the cook, and Chanda, his assistant, run from the kitchen to welcome me back. But Mark is flying a patrol and won't return until much later. Swallowing my disappointment, I plan a special dinner for us. Mumanga, who thinks Mark and I never eat enough, pitches in.

Most of the open kitchen area has been without a roof since Survivor tried to eat it. The little wood stove, whose legs collapsed under the roof, is now set on blocks under a small, round, thatched roof. In no time Mumanga has the fire going. Smoke belches from the tall chimney with its cocked tin hat.

"Mumanga, you make a cake, okay? I'll bake a tuna pie — Mark's favorite. Chanda, please set the table in the little din-

* The Owens Foundation for Wildlife Conservation, which depends on donations from the public, supports our work. Tax-deductible contributions may be made to the foundation at Box 53396, Atlanta, Georgia 30355. We send periodic newsletters to contributors. Our other major sponsor is the Frankfurt Zoological Society of Germany, 6 Alfred-Brehm Platz, Frankfurt am Main 1, Germany.

ing cottage. Mark and I haven't had a real dinner together in months."

As we dash around the kitchen, up to our elbows in pastry and batter, Chanda and Mumanga tell me about the wildlife they have seen near camp. The zebra without stripes grazes across the river every so often, and the small herd of buffalo come into camp every morning.

Late in the afternoon the plane zooms over us as Mark approaches the airstrip. Everything is ready: Mumanga is icing the cake, the pie is baking, the table in the dining cottage is laid with candles and pottery dishes.

His eyes bloodshot, his face drawn, Mark steps out of the truck at the workshop. He hugs me briefly, and before we have taken two steps he announces that he must go straight back to the airstrip. Musakanya has passed the word that poachers may be heading into the park from the north, to a hilly area where Mark does not fly often.

Standing in the n'saka, Mark wolfs down a peanut butter sandwich and chases it with thick black coffee. I don't mention the special dinner I have made, or suggest that he take a rest, because I know he won't listen. According to the doctors in Lusaka, his collapse was brought on by stress and fatigue, by parasites and a virus, and by too much caffeine. All the same, he has ignored the warnings to slow down.

Sitting on the low stone wall of the n'saka, Mark drinks more coffee as we talk. The cherry bombs have cleared the park of poachers more effectively than we could have imagined, but we know this cannot last. Sooner or later the poachers will realize that the firecrackers are harmless; sooner or later Mark will be shot down or arrested by a corrupt official.

We have to try again to get the game guards to do their jobs. Our last hope, we have decided, is to find a tough, committed Zambian to serve as their unit leader. The encouragement, motivation, and leadership that the Mano scouts so desperately need should not come from us, but from one of them. We need a Zambian as crazed about elephants as we are.

Only two men can help us: Luke Daka and Akim Mwenya.

Together they are in charge of all the scouts, wardens, and administrators of National Parks. It was Daka who fell in love with Survivor when he met him at our camp. We will fly to Lusaka in a few days to meet with them again.

Mark, not noticing the pie or cake in the kitchen boma, quickly kisses me good-bye and drives back to the airstrip. So that I won't hurt Mumanga's feelings, I follow through with the evening meal, sitting at the table alone, eating a dinner meant for two. As the sun sets, I hear the plane take off in the distance and disappear among the hills of the scarp. I have never felt so lonely. After dinner I pack fresh supplies for my camp, give some cake and pie to Mumanga and Chanda, and ask them to keep the rest warm for Mark's return. Then I climb into my truck for the three-hour drive to the Luangwa. The half-moon will be my companion.

o o o

Acrid smells and shrill city noises filter through the open windows of Electra House, office of the Ministry of Tourism in Lusaka. Mark and I sit at a conference table with nine men from the Zambian government: Luke Daka, permanent secretary of the ministry, the director and the deputy director of National Parks and Wildlife Services, several other high-level officials from that department, and a few representatives of the Anticorruption Commission whom we have invited. We have explained the poaching problems to all of these men many times during meetings in Lusaka. But never have we met all together.

"Mr. Daka," Mark begins, "you will remember Survivor, the elephant who came to our camp when you were there. I'm afraid I have to tell you that poachers came to Marula-Puku and shot into his group, killing at least one elephant. We believe it was Survivor."

Daka frowns. "No! That's terrible!"

"It gets worse. I found the poachers' camp from the air," Mark went on, "but the game guards refused to go after them."

"This is outrageous! How can game guards refuse to take action in such an emergency?" Several men squirm in their seats.

"Sir, this is just the beginning." Mark talks for twenty minutes, telling about the radio and gun licenses not being granted, about scouts seldom patrolling, about corruption in Mpika, about National Parks officers — including the warden — who deal in ivory, skins, and meat. He describes how he has been shot at again and again while aiding the scouts, and how an official has accused him of buying black-market military weapons.

"But," Daka stammers, "this is ludicrous. Who has made these charges against you? And why two years to get radio licenses — they should be ready in two or three days." He glares at the men around him.

"We believe," Mark continues, "that people with poaching interests in North Luangwa are trying to block everything we do. They want to keep us from getting firearms and radios so the scouts can't protect the park."

Mark pauses to let the message sink in. "Sir, North Luangwa is one of the most beautiful parks in Africa — one that could, along with South Luangwa, bring millions of dollars into Zambia through tourism. But I can only say that we cannot go on like this. We have to know that this government is going to support us, or at least not undermine us."

Daka stares at the table, twisting the pen in his hand. "Mark, Delia," he finally says, "after being at your camp, I can see that your project is the only hope for this park. We WILL support you in every way."

Mark and I have heard this before; we are skeptical. But then Akim Mwenya continues, leveling his eyes at his junior officers from National Parks, "Tomorrow I want someone from the department to fly to Ndola to get these radio licenses. I also want the licenses to bring these guns into the country right away, so that the scouts can be armed properly. And new personnel should be sent to the Mano Unit."

The director offers to immediately send twenty scouts with special military training to Mano and build a new camp for them. North Luangwa will be given top priority.

"Sir, I would like to make one more request," Mark interjects. "The unit desperately needs a dynamic new leader. Please send us the best man you have."

We thank them all, especially Mr. Daka, but after four years of false hopes and disappointments we leave feeling more anxious than optimistic.

"Wonder how many hours they'll give us to pack up and get out of their country," Mark jokes as we step out onto the street. Walking through the jostling crowd, I do not laugh.

o o o

Up before dawn, I count the hippos from my camp before driving up the scarp mountains to meet Mark near Mano. Since our meeting in Lusaka a few weeks ago, three senior government officials have been suspended, pending investigation of charges that they were smuggling ivory to Swaziland. Our radio licenses have been approved and fresh scouts assigned to North Luangwa. A new unit leader has arrived in Mano, and Mark and I are to meet him today.

Driving along different tracks, Mark and I pull up at the Mwaleshi River near Mano at about the same time. We climb out of our trucks to inspect the new pole bridge we had built with local labor, then drive along the new Mano-to-Mukungule road to the recently completed airstrip. To our astonishment, a squadron of game guards dressed in full uniform marches double time along the strip toward us. They halt, turn smartly in place, honor us with a crisp military salute, then march off again. At the head of their column is a lanky, young Zambian wearing a proud smile. Incredulous, Mark and I look at each other. Can this be the Mano scouts drilling?

In Mark's truck we drive to the main camp. The children swarm around us as we step down from the Cruiser, then ask us to wait in the crumbling n'saka while they fetch the new unit leader. Ten of them race off to get him, while the other children ask for stories from the Luangwa Lion puppet. Mano still looks more like a refugee camp than an official game scout unit headquarters. The cracked walls of the old huts have simply been plastered with new mud that will soon crack in the sun.

Ten minutes later, the children dash back across the field ahead of the young man and the column of scouts. He orders his drill team to about-face, they march back down the field, and he dismisses them. Sweating heavily from the march, but crisp in his new green uniform, he walks briskly toward us. I stare at him. This is it, as far as I am concerned. If this man cannot bring Mano Unit under control, the project will be forced to recruit its own scouts. If that doesn't work, we will find another wilderness to save.

His handshake is firm and he looks me straight in the eye, his gaze steady and confident as he introduces himself. "I'm Kotela Mukendwa. I've heard all about you. We have much to discuss. Please sit down."

As Mark and I brief Kotela on the poaching in North Luangwa and our problems with the scouts, he nods his head knowingly. He intends to turn these undisciplined guards into a crack military-style unit, he says. We promise whatever he needs within reason to get started; if he does the job, we will consider requests beyond reason.

Talking so fast we can hardly understand him, Kotela presents a neatly prepared list of his needs: use of a truck to capture known poachers in the villages, fuel, more guns, ammo, and food for patrol. He will drill the men every day, instruct them in military tactics, order them to be dressed and ready for patrol at all times. He has sketched in detail his plans for an office, jail, armory, and storage complex for Mano. Almost dazed by his competence and determination, we agree to get him virtually everything on his list.

Mark stands, offering Kotela his hand. "Let's do it!" And he smiles his first real smile in a long while.

o o o

Mark and I set up a little camp near the waterfall, across the Mwaleshi River from the scout camp, so that we can help Kotela as much as possible. Using our trucks and funds for fuel and food, he provisions Mano with supplies and lectures the men on patrol tactics. He hires informants and prepares the scouts to go on village sweeps, in which they will raid poachers' homes in the middle of the night and arrest them. Meanwhile, we install new solar-powered radios at the various scout camps, organize farmers to grow food for the scouts, and in general try to improve their living conditions. Using money we have raised in the United

States, we purchase two new tractors and trailers to supply the camps, and a grader to make better roads and airstrips for the budding tourist industry.

An old man from a nearby village knows how to make proper adobe houses, so we hire him to build new cottages for the scouts and their families. Using beautiful earth shades of rust, red, brown, and green, the scouts' wives paint the new mud houses with striking geometric patterns. We hire a crew of sixty villagers to improve the track into the park.

The special military-trained scouts promised by the director show up and settle into the new houses. Every dawn Kotela can be heard barking orders as he drills them on the airstrip. Looking neat in their new uniforms, the scouts salute their officers smartly. The old beer pot — formerly the center of activity — has disappeared. Mano Camp, the once-dreary den for bedraggled scouts, is pulsing with new energy.

We hire two English lads, Ian Spincer and Edward North, who are fresh out of the University of Reading. They are unsuspecting and ready for anything, so we station them at Mano to help Kotela organize camp logistics and a law-enforcement program. Ian, a graduate in agriculture, begins a farm at Mano that produces vegetables, rabbits, and poultry for the scouts and their families. He installs a mill to grind their corn for mealie-meal, and supervises the delivery of all foodstuffs to the remote camps by tractor and trailer. Edward sets up a firearm training program for the scouts, issuing the new guns we have imported from the States. Both he and Ian patrol with them, evaluating their field performance. We purchase more camping equipment and a truck for the scouts, and Simbeye moves to Mano as their official driver. He is also in charge of training an auxiliary force of local villagers who will aid the scouts.

Kotela, Ian, and Edward organize the unit into seven squads and devise regular schedules for patrols into the park. They employ a regiment of porters — most of whom previously worked for the poachers — to carry food and supplies to scouts in the park. Now the men can patrol for as long as three weeks, covering much larger areas, without Mark's having to fly dangerous resupply missions. For the first time ever, there are at least some scouts in North Luangwa National Park at all times. Mark creates special units for those who perform well, and they are issued extra equipment — new guns, jungle knives, binoculars, compasses.

Everything is in place; but we have yet to capture poachers.

One afternoon, after supervising the road crew all day, Mark and I drive into the main camp at about four in the afternoon. The place seems deserted, and then we see that all the wives, children, and men of Mano are standing around the n'saka. The scouts have returned from their first village sweep, and the n'saka is crowded with fourteen meat poachers, a pile of illegal guns stacked against a tree outside.

The sweep continues for four more days. The scouts raid villages all night — bursting into poachers' huts while they sleep — and drive back to Mano in the morning, their truck loaded with suspects and illegal guns. The poachers have acted with impunity for so long that they are caught off guard. Moving swiftly, Kotela and his men capture dozens before the word gets out that the scouts are back in business.

The old game guards, like Island Zulu and Tapa, walk taller now and salute us with pride. No longer full of excuses and complaints, they tell us wildly exaggerated tales of capturing notorious poachers. Nelson Mumba, still wearing his red bandana and refusing to patrol, and Patrick Mubuka, the scout who shot elephants in the park, are put in the back of a truck, driven to Mpika, and dumped at the warden's door, never to return to Mano.

While Kotela and his scouts patrol the park and raid the villages, Mark continues to terrorize the poachers from the air. Suddenly Mano has become the number one unit in Zambia, capturing more poachers than any other.

The five camps of Mano Unit are only effective on the western border of the park. To protect its northern flank, we send Edward North to assist the scouts and villagers of Fulaza. We hire an Alaskan bush pilot, Larry Campbell, to rebuild the scout camp in Nabwalya, south of the park, as well as to solve some of the problems in the village.

Dramatic as the changes are, they are just a beginning. Although Kotela is having great success rounding up many of the commercial meat poachers, most of the big ivory poachers in villages like Mwamfushi are still eluding the scouts with their reputed powers of invisible juju. The park covers twenty-four hundred square miles, and only twelve game guards patrol at a time. Experienced men such as Chikilinti and Chanda Seven have little trouble avoiding them. Mark chases them out of the park, but they come right back. The scouts capture one; the magistrate lets him go. Somehow we have to get these men of Mwamfushi Village.

23
Mwamfushi Village

DELIA

Dear Directors
North Luangwa Conservation Project

My name is Steven Nsofwa and I am thirteen years old
and I student at Mukungule Primary School. Please, I
wanting thank Madam and Sir for the things you have
done for our village. Now since you are coming the maize
mill makes our mealie-meal, the shop sells our soap, the
school now has a map of the world. We the students love
to play the games about the elephants and draw pictures
of the lions. When we grow up we will chase the poachers
from our village so that there will always be the animals in
the rivers. For now we are not too big to chase them.
Please we want to thank you.

Yours,
STEVEN NSOFWA

o o o

DODGING DEEP GULLIES and ruts, Mark steers the truck toward Mwamfushi Village. On the seat between us are Mark's
leather flight bag and my briefcase. In Mark's flight bag —
jammed among the aeronautical maps and papers — is his 9-mm
pistol; in my satchel is a .38 caliber revolver. As we drive along, we
scan the tall, waving grass along the track, searching for any sign
of an ambush.

We have been warned not to go to Mwamfushi. Even Kotela
tells us it is too dangerous. The men in this village have tried to
kill us, they've shot at Musakanya's house with automatic weapons,
and they have poisoned Jealous and several other people who work
for us. But if we can't stop the poachers of Mwamfushi, there is
no hope for saving the elephants of North Luangwa. We have sent
word to the headman of Mwamfushi that we would like to meet

with his villagers this Saturday morning at the schoolhouse. We are gambling that the poachers won't risk attacking us in broad daylight, with other people around.

Stopping along the way at scattered huts and bomas, we pick up Chief Chikwanda and his retainers. Also with us is Max Saili, who is in charge of our project's work in the villages. When we drive up to the clay-brick school, our pickup splattered with mud and loaded with dust-coated officialdom, thirty villagers are waiting in the barren schoolyard. A few old men, dressed in patched trousers and ragged shirts, shake our hands; but most people just stare. Clutching our cases tightly, we make our way through the crowd.

As we enter the schoolroom, Mark discreetly points to a man who is a big buffalo poacher. None of the ivory poachers — Chanda Seven, Chikilinti, Bernard Mutondo, Simu Chimba, or Mpundu Katongo — have come, but I am glad this man is here; if he will exchange his weapon for a job, that will set an example for others to follow. Saili, Mark, and I sit in chairs that have been arranged at the front of the room; the villagers sit on the students' benches. We all stand as Saili leads us in the singing of Zambia's national anthem. Its soft, mournful melody drifts out of the cracked windows and across the untended fields.

Standing to give the opening speech, Honorable Chief Chikwanda is wearing a T-shirt we have given him earlier. The print of a large elephant adorns the back, and whenever he repeats the word "elephant" in his speech — which is very often indeed — he whirls around, swaying back and forth, presenting the dancing elephant to his audience. Shouting and strutting like an evangelistic preacher, he reminds his people over and over that wildlife is a valuable resource, that it is the heritage of his people, and that poachers are sacrificing their children's future. He ends with a stern warning that poaching in his chiefdom must cease immediately. It is quite a performance, but it would be much more impressive if everyone in the room did not know that Chikwanda himself has been charged with elephant poaching more than once. At this very moment, in fact, he is appealing one of his convictions.

Saili introduces us, and as we speak he translates after every few sentences. We begin by explaining that we are not here to arrest anybody; that if, by chance, any poachers are present (we know that probably 85 percent of the people here are involved), we will give them jobs if they turn in their weapons. We know they are poaching because they need food and work, and we are here to help them find alternatives. It is a long, halting business with Saili translating, but finally we exhaust our supply of convincing and worn-out statements. There are murmurs of approval from some of the older, toothless crowd, but several of the younger men argue heatedly among themselves.

A youth in a faded pink shirt stands up and says in English, "This what you say may be a good thing, but there are many men in this village — not myself, of course (laughter) — who have to poach. You maybe to hire some of them, but you cannot hire us all. There are plenty." Several people nod or shout in agreement.

Another young man, no more than sixteen, raises his hand. "A friend I have who carry for the poachers. What will he do if poachers go from this place?"

An elder stands. "That is what the Owenses are saying, they will help us find other work."

"What are you saying, old man?" shouts the man in pink. "It is you who gives your daughters to the poachers for some little meat!" The room erupts as men shake their fists and shout at one another in Chibemba. One elder stomps out of the room. Raising his hands and speaking Chibemba, Saili eventually restores order.

When the crowd is calm, Mark speaks again. "We know this will not be easy. But your whole village depends on poaching, which is a crime. Your own children are told to be carriers, which is illegal and dangerous. And you know better than I that if you continue shooting the animals they will all be finished. Think about how far your father had to walk before finding an elephant, buffalo, or hippo to shoot. Your grandfathers and fathers could get meat right around your village; now you must walk forty or fifty miles to find any animals. Gentlemen, there are so many people and so few animals now, that they cannot feed you anymore. If you keep killing them, soon they will be gone. So you must find ways to make money by keeping them alive. And we are here to help." Saili translates.

Again a murmur of discontent rattles around the room. Mark and I look at each other, disappointed; things are not going as we'd hoped.

Then Mark points to the buffalo poacher in the back of the room. "I'll start with this man. I know you, I have seen you in the park. I'll give you a good job right now. Come and see us after the meeting." The crowd erupts in laughter. The man grins at his friends, but says nothing.

After forty-five minutes of shouting and arguments, I pass around paper and pens and ask those interested in working for us to write down their names and skills. A few young men walk hurriedly from the room, probably nervous about revealing their identities. But most of those who can write scribble their names, and the names of their illiterate friends.

Seven of the men — including the man in the pink shirt — claim to be trained carpenters. They are out of work because the tools from their cooperative were stolen and they cannot afford new ones. One man is a tailor; several know how to make and lay bricks; two are licensed drivers. The village is full of men who, if given a chance, could earn an honest living. Mark and I whisper quietly together, making some quick decisions.

"We'll start right now to help your village," I say. "First, we'll lend you cash to buy the tools, wood, and hardware necessary to open a carpentry shop. When you are making furniture and bringing in money, you can gradually repay us, so that we can help others."

"But remember," Mark interrupts, "this is not a gift. In exchange, you must stop poaching and you must chase the big poachers from your village. Do you agree?"

Almost everyone, including the man in the pink shirt, nods.

"Also," Mark adds, "we'll hire a labor crew to rebuild the road between your village and Mpika. Musakanya, who was once a poacher from your village but who now works for us, will be the supervisor." Before he has finished speaking, men are scrambling around Musakanya, asking to be chosen for the crew.

We invite them to apply for loans to start other cottage industries — cobbler shops, peanut presses for making cooking oil —

but caution that we cannot grant everyone a loan. Priority will be given to the businesses that employ the most people or produce food or some essential community service.

By now it is well after lunch. Almost no one has left the school-room and many others have come in. Exhausted, we close the meeting with another singing of the national anthem and make our way through the crowd of people pressing around us to ask for help. A small woman, dressed in a red chitenge, grabs my arm. "Madam," she says, "you have forgotten the women."

I stare at her for a moment, feeling embarrassed. "I promise, I haven't forgotten you. It's just that the men do the poaching, so we've offered them jobs first."

"You do not understand," she continues. "The women are very dangerous in this village. They do all the work — farming, house building, cooking, washing — while the men sit under the trees. So it is the women who say to their husbands, 'Why don't you go to the park and get some bush meat?' It is the women who tell their children to leave school to work as carriers for the poachers."

I have not heard this before, but I can see the truth in it. "How can we help?" I ask. Standing in the dusty schoolyard, surrounded by chattering villagers, scampering children, and the ubiquitous scratching chickens, the women and I make a deal to set up a sewing shop for them.

When at last we reach our truck, still surrounded by would-be converts, I see the buffalo poacher walking quickly around the school toward the fields. He has not come to us for a job. This saddens me, but I feel that otherwise the meeting has been a huge success.

Not wanting to lose momentum, we immediately send Tom and Wanda Canon, our project volunteers, to Lusaka to buy carpentry tools, a manual sewing machine, a grinding mill, and tools for the road crew.

A few weeks after our first meeting we drive back to Mwamfushi with Tom, Wanda, and Saili. Eighty men and women are crammed into the schoolhouse, chattering excitedly as they wait for us. Small children line the walls outside, chinning up to poke their faces through the windows. The villagers clap, sing, and ululate a high-pitched, spiritual melody as we present the sewing machine to the women and the tools to the men. We clap along, then open a discussion about other industries that can be started in the village and how poaching can be stopped.

While Saili is translating, a young man slips through the doorway into the schoolroom and hands Mark a scrap of paper. He reads it, grins, and hands it to me. Scribbled in a wavery block print are the words, "I want to joine yu. I give my weapon. Please to met me in shed behind the school before you going. Cum alone." It is signed "Chanda Seven."

o o o

"Mark, you can't just walk out there by yourself," I whisper. The meeting is finished and almost everyone has filed outside. We are standing alone in a corner of the schoolroom.

"I'll be careful."

In the schoolyard Saili, the Canons, and I continue talking to the villagers. Meanwhile Mark, his hand in his flight bag, disappears around the corner of the school, heading for a crumbling mud-brick shed that stands on the edge of a maize patch. A wooden door dangles on one hinge, hiding the dark interior. Mark stands against the wall, listening for sounds from inside, then kicks the door open, and peers in. Gradually his eyes adjust to the darkness, and the outline of a man takes form. He is standing in the shadows behind a crude counter, his hands on its top. Near his fingers — a twitch away — lies an AK-47.

"You are Chanda Seven?" Mark asks, his eyes riveted on the rifle. "What do you want?"

"I'll give you my weapon if you give me work."

"The first thing you must do is step back from the counter." Chanda Seven moves deeper into the gloom. Mark steps quickly inside, takes the rifle by its barrel, and leans it against the wall.

"I can give you a job. But how do I know I can trust you? You have tried to kill me and my wife, you have shot many elephants, you have shot at Musakanya's house, you have poisoned Jealous."

"I do not know how you can trust a man such as me. It can be only that I give you my weapon. Then I am unarmed."

"It is not enough to give me your weapon. You might come to work for me, learn our routine, then lead other poachers to kill me. To work for me, you must first prove I can trust you. Help me capture the other poachers from Mwamfushi — Simu Chimba, Chikilinti, Mpundu Katongo, and Bernard Mutondo — and I will give you a very good job. You must find out when and where they plan to poach, and send a message to me through Musakanya. Do you agree?"

"Eh, Mukwai, yes."

Mark puts the AK in a gunnysack and shakes hands with one of the most notorious poachers in Mwamfushi. Walking back across the crowded schoolyard, Mark gives me a thumbs-up salute.

I smile. One of the big guys taken — without a shot fired, without a dangerous night flight.

24

Sharing the Same Season

DELIA

The struggle for any dream
is always worth the effort,
for in the struggle lies its strength,
    and fulfillment
        toward the changing seasons
        of ourselves.

— WALTER RINDER

o o o

IN MAY 1991 the marula fruits lie on the ground bursting with a
honey-sweet fragrance, and my mind turns once again to Survivor.
Did Mpundu Katongo, Chanda Seven, and Chikilinti kill him or
did he escape? Am I foolish to hope that he will wander into camp
on his return from the plains? Mark continues to fly antipoaching
patrols and count elephants; I spend most of the time at my river
camp, encouraging scouts, or working in Mwamfushi. But I never
drive through the hills near Marula-Puku without thinking of Survivor, without searching the ridges and valleys for an elephant with
small tusks and a hole in his left ear.

One afternoon a few weeks after our visit to Mwamfushi, I am
sitting on the red, dusty ground with the children of Mano, drawing pictures of elephants. The surrounding forest trills with the
songs of lovebirds and wild parrots, and the encampment hums
with the steady sounds of the women cooking and washing outside
their mud huts. The drone of an engine drifts through the trees:
Mark is grading the Mano airstrip with one of the new tractors.

A plume of dust rising behind it, Mano's truck roars into camp
with ten scouts in the back, singing and holding their thumbs up.
We have never heard the scouts sing before; the children and I

jump up and run to meet them. A ragged, dusty man sits in the back of the truck, handcuffed, his head hanging.

"We have captured Simu Chimba, Madam! Look how small he is! A big poacher like this, how can he be so small?"

"Well done!" I congratulate them, thinking, "Two down, three to go."

o o o

With Mark patrolling in the air, the scouts and our team busy on the ground, we haven't seen a poached elephant in eight months. Our work to win over the people of Mwamfushi continues, although in a painfully halting way. Everything needed to supply the budding cottage industries, from sewing needles to hacksaw blades, must be trucked all the way from Lusaka or imported from abroad. But already there is a change in the village; it seems to whistle. Musakanya and his road crew have built a broad, smooth track from Mpika to the Mwamfushi; the mill is grinding mealie-meal; the sewing shop is manufacturing new uniforms for the game guards; the carpenters have produced a desk. One person talks of making candles, another of making soap, a third of developing a fish farm. We have started a conservation education program in the school, and posters of elephants, lions, and leopards — courtesy of the Dallas Zoo — give splashes of pride and color to the clay-brick walls.

With a new sense of authority and hope, the people of the village have been spying on Chikilinti, Mpundu Katongo, Simu Chimba, and the other commercial poachers, and reporting their plans to Musakanya. By harassing them with insults and threats, the Mwamfushi vigilantes have forced the poachers out of the village into the bush, where they hide out in grass hovels, afraid even to have a campfire at night. Their power over the villagers has been broken. Chanda Seven and Musakanya, the two ex-poachers now working for us in Mwamfushi, are shining examples for the unconverted. Both have supervisory positions — Musakanya with the road crew, Chanda Seven on a farm — and are pulling down good salaries without breaking the law.

One lazy afternoon, while all of Africa is nodding in the heat, Mpundu Katongo wanders down from the hills toward Chanda Seven's hut. He is planning a major poaching expedition into North Luangwa and hopes to persuade Chanda Seven to join him. Mpundu has heard all the nonsense about fish farms and grinding mills, but does not believe any of it will pay as well as poaching. Although he is short, he is very strong; and unlike some of the other poachers he is not frightened by the villagers.

Chanda Seven sees his old friend across the field and invites him into his mud-wattle hut for beer. They shake hands in the neatly swept yard, and Chanda Seven steps aside, allowing his guest into the hut. As they duck through the tiny doorway, Chanda shoves Katongo against the far wall, jumps back outside, and locks the door. Katongo shouts and bangs on the puny door, threatening to break out. Grabbing an ax, Chanda Seven swears he will chop Mpundu into tiny bits and feed him to the hyenas if he tries to escape. Drawn to the commotion, vigilantes armed with hoes and rakes race through the maize patches to help Seven contain his captive, shouting insults at him through the door of the hut.

When Musakanya sees what has happened, he runs five miles to Mpika to inform the game guards. It takes some doing to convince the Mpika scouts to come, but eventually, using one of our trucks, they drive to Mwamfushi and arrest Mpundu Katongo and take him to Mpika. Early the next day, even before we have heard of the arrest, Max Saili radios us to say that he has heard the warden is going to release Katongo.

"Oh no he's not!" Mark shouts over the mike. "Don't let anything happen! I'll be in Mpika in thirty minutes." Mark jumps into the plane and flies to Mpika, where Saili meets him at the airstrip. No doubt a bribe has changed hands, but this time it won't work.

At ten o'clock in the morning Mark roars up to Mpondo's Roadside Bar and jams on the brakes, a cloud of dust swirling behind him. He marches inside the ramshackle building and looks around the dimly lit room. Warden Mulenga is sitting alone at a table with eight empty beer bottles. He glances in Mark's direction and then stares blearily at the wall.

"I heard you're going to release Mpundu Katongo?" Mark demands.

"Insufficient evidence," the warden slurs.

"You know this man is a poacher. If you don't have enough evidence, we do. If you release him, I'm going to make a big stink at the ministry."

The drunken warden, mumbling something unintelligible, tears a scrap from a brown paper bag and scribbles his authorization for Katongo to be taken to Mano for questioning. He agrees not to drop the charges against Katongo as long as we hire him. Mpundu will be in custody of Kotela and the Mano scouts.

"Thanks, warden, have another beer." Mark slaps some kwacha notes on the table and walks out.

Handcuffed and guarded by two scouts, the poacher is lifted into the plane. Mark flies him to Mano airstrip, where the scouts and I meet him. They swarm around the plane, and when Mpundu is handed down to them, they march off toward camp, pushing and shoving the shackled captive. Once in the n'saka the scouts hold Mpundu down on the ground and paint blue and yellow lines on his face — a juju that removes all power from the poacher. Hands and feet bound, he sits in the center of the n'saka while the scouts, wives, and children take turns humiliating him. One scout demands that Mpundu act like a chicken. Hobbling around the camp, he scratches the dust with his bare feet and flaps his arms as best he can. Every few steps he topples over and falls hard to the ground, and everyone laughs.

When the scouts at last tire of the ritual, Kotela tells us that we can interview the prisoner. We set up the video camera and I stare into the eyes of Mpundu Katongo. He is a short, stocky man with a bulldog's face. Stripped of his juju, he sits quietly staring into the dirt and confesses that he has shot more than seventy-five elephants, hundreds of buffalo, and too many puku to count. Speaking clearly to the camera, he admits shooting at the airplane with semiautomatic weapons on many occasions. In vivid detail he describes how he and the others planned to attack our camp but abandoned the scheme when they saw the game guards. He tells us that it was he, Chanda Seven, and Chikilinti who shot the elephants near Marula-Puku.

We bring Katongo's family to Mano and give them a house. We hire him to lead the scouts on patrols, using his knowledge of routes and hunting areas to help them ambush and apprehend other poachers. Later we also hire Bernard Mutondo, the poacher who killed a game scout and wounded three others south of the park and nevertheless was released by the magistrate. Now three of the five men who tried to kill us are on our payroll.

o o o

Stepping quietly through the undergrowth, I walk away from Nyama Zamara lagoon toward my river camp. A large male waterbuck, standing in the tall grass, swings his head in my direction. I don't move. He looks at me for a moment, then continues to graze. I walk on toward the beach, where sixty hippos are sprawled on the damp sand.

Unexpectedly I hear the sound of our airplane to the south; Mark must be patrolling the Mwaleshi River. As always when I hear the plane, I take the walkie-talkie from my backpack and switch it on in case he calls me. Although he often patrols in this area, he has never visited my camp. The drone of the engine grows louder.

"Brown Hyena, this is Sand Panther." Mark's voice crackles over the radio. "Do you read me?"

I can see the plane swooping low above the trees at the river's bend. "Roger, Sand Panther. Go ahead."

"Hi, Boo! Want some company for dinner tonight?"

"Roger, Sand Panther. That'll be fine. As long as you understand that dinner at my camp is a jacket-and-tie affair," I joke. "And don't forget the chocolates."

"Of course," Mark laughs over the air. "I'll fly back late this afternoon. Please pick me up at the airstrip on the plains where we darted Bouncer."

"Roger. See you then. Brown Hyena clear." I zip the radio into my backpack and run through the trees toward camp, avoiding the beach so that I won't frighten the hippos. What in the world will I cook for dinner?

In my little grass hut I fling open the blue storage trunks and rummage through the tins looking for something special. I decide on packaged onion soup, pasta with canned mushroom and cheese sauce, green beans, and cookies. I set the table with a chitenge patterned with swirls of greens, blues, and reds, candles set in snail shells, and yellow enamel cups and plates. At the edge of camp I pick tiny blue wildflowers and place them on the table in an empty peanut butter jar. Apparently sensing my excitement, the baboons scramble into the ebony tree over the hut and lean from the branches to watch me.

Folding a sheet of paper in half and decorating it with sketches of hippos and crocodiles, I write out a menu for the evening:

Appetizer
Smoked oysters Zamara
Soup
Luangwa onion soup with garlic croutons
Entrée
Lagoon pasta with mushrooms and cheese
Green beans Lubonga
Dessert
Moon cookies

I have no decent clothes at my bush camp. Determined to look as spiffy as possible, I make a skirt of another chitenge and wear it with a white blouse, sandals, and my favorite Bushman earrings. In the late afternoon I drive to the grass airstrip on a flat open plain about four miles from the river. During the twenty-minute trip I pass herds of wildebeests, zebras, and buffalo grazing along the track. I am a mile from the strip when Mark swoops down in the plane right above the truck. Perfect timing. He will land just before I arrive.

By the time I turn onto the dusty strip, the plane is parked at the far end. But driving toward it, I cannot see my pilot. Strange — he should be out of the plane by now, tying it down for the night. Driving closer, I am more and more puzzled; the plane stands deserted. I park next to it, step out, and look around.

Suddenly Mark struts from behind the fuselage where he has been hiding. Below the belt he is dressed as usual — khaki shorts, bush socks, and desert boots. But above the waist he is wearing a smart blue blazer, a white dress shirt, and a tie. A wiry bunch of wildflowers hides his devilish grin, and a bottle of champagne is tucked under his arm.

"Bonjour, madame. Sand Panther at your service," he says with a sweeping bow.

On the bank overlooking the beach, we nibble smoked oysters and sip champagne, watching the sun and the hippos sink into the broad river. Later, with moonlight streaming on the white sand, the hippos stagger from the water and stroll just below our din-nertable perched on the edge of the bank. My little grass hut glows with soft candle and lantern light. Whispering and laughing softly, we search the sky for shooting stars — and count six.

Later, on our bedrolls in the tent, I snatch up my pillow and look for chocolates. A long time ago it was Mark's favorite hiding place for special surprises, including chocolates. But there is noth-ing. Hiding my disappointment, I slide between the sheets and he pulls me close — laughing. Then my feet touch something at the far end of the covers.

"Mark, what have you done now?" Peeling back the blanket, I discover chocolate bars — twenty-five of them — lined up at the tips of my toes.

o o o

It is the dry season again. The grasses are tired now, having made seeds in every imaginable shape and array. They lie on their sides, resting their heads near the ground. Eventually the wildfires will consume them, sending their last drops of life to the clouds, which in turn will rain down on the saplings of distant hills. Or, if the fires do not come, the grasses will return to the soil, giving their souls to new hopeful seeds. Either way, we will see them again. Even the colors are weary, having burned themselves out with brilliant golds and reds, then fading to the pale hue of straw. Life is taking a breather, and the year itself must rest.

It is the dry season again. It comes every year. But I know now that the life-giving rains will return. Just as there is an end to winter, there is an end to drought. The secret is to *live* in every season. The Kalahari taught me this, and like the desert I want to sing in the dry season and dance in the rains.

Since the poachers were stopped before they could set their wildfires, for once there is grass for the animals to eat, long into the dry season. Meandering lines of elephants drift across the valley floor, feeding on the savannas. In former years, by March or April they would have fled into the protective hills of the scarp, abandoning the superior forage of the valley grasses for the fibrous bark of trees in the miombo woodlands. But now Long Ear, Misty, Marula, and their young feed along the open floodplains of the Mwaleshi, where the tall elephant grass waves in the gentle breeze.

At the airstrip one morning, Kasokola and Mwamba, who are guarding the plane, tell us they have seen two elephants in the small valley near Khaya Stream below the hill. One, they say, has tusks; one does not. For the next few days we look for them but see nothing. Several mornings later, we are passing Hippo Pool on our way back to camp when a bull elephant without tusks appears in the tall grass near the bank. It is Cheers. Every day after that we find him feeding on the long grasses across the Lubonga near camp, or on fruits near the airstrip; but he will not come closer. Always there is another elephant with him, standing deep in the trees, so that we can never get a good look at him. Like a gray shadow, he always slips silently away.

Late one afternoon, sitting quietly by the river, Mark and I hear a rustling on the far bank. Slowly, an elephant plods from the tall reeds of a dambo on the other side and stands looking at the marula trees behind us. I put my hands to my face in disbelief. The elephant has tusks as long as Mark's arms and a tiny hole in his left ear. It is Survivor.

For several minutes he watches us with eyes that have seen too much. He lifts his trunk high into the air in our direction. But we are not fooled; this is not a greeting. It is the marula fruits that he smells and the marula fruits that he wants, not contact with us. He is not thanking us for being here, or blaming us for not doing a better job of protecting him. He·just wants to eat these fruits, wander these hills, and live with his own kind. It is not too much for his kind to ask, or for our kind to give.

He gives us a wide berth as he passes on the far bank, not coming nearly as close to us as he did last year. He lumbers to the river, touching the water with his trunk, lifting it to his mouth. He stands for long moments, looking at us, then glides silently along the sandbar. It is said that elephants do not forget; perhaps they do forgive.

"What are we going to do now?" Mark whispers to me.

"What do you mean?"

"You've always said we'd go home when the elephants could come to the river and drink in peace."

I look along the Lubonga, where it gently sweeps through the high banks and rocky shoals. Five puku lie in the cool sand near a pair of Egyptian geese.

And I answer, "We are home."

Return to Deception Valley

DELIA

The land has been hurt. Misuse is not to be
excused, and its ill effects will long be felt.
But nature will not be eliminated, even here.
Rain, moss, and time apply their healing bandage,
and the injured land at last recovers.
Nature is evergreen, after all.
— ROBERT MICHAEL PYLE

o o o

DURING ALL OUR YEARS in Luangwa we never forgot the Kalahari. Whenever we saw the Serendipity Pride, we thought of the
desert lions; whenever it rained at Marula-Puku, we wondered if
East Dune was still dry; whenever there was a full moon, we
longed for Deception Valley.

In 1988, while we were in North Luangwa, we received a letter
reporting that a special commission in Botswana had decided the
fate of the Central Kalahari Game Reserve. Among the alternatives it had discussed was dissolving the lower two-thirds of the
reserve so that it could be used for commercial cattle ranching. In
the end, under scrutiny by international conservation agencies, the
commission voted to keep the entire reserve intact. In addition, it
adopted many of the environmental recommendations we had
made before all the controversy. These included taking down certain fences to open a corridor for migratory species such as wildebeests and hartebeests. Of course, by then most of the wildebeest
population — more than a quarter of a million animals, as well as
tens of thousands of other desert antelope — had perished. But
perhaps if good rains return to the Kalahari, this harsh but resilient
land will once again provide the miles of golden grass necessary to

bring back these populations. Man has dealt the Kalahari a staggering blow, but deserts know all about rebirth.

Inasmuch as our Prohibited Immigrant status had been reversed, once the decision was made to save the reserve, nothing could keep us away from the moody dunes and ancient river valleys that had been our home for seven years. We planned an expedition to clean up our campsite, to remove every trace of our having lived there; to search for the lions we had radio collared before our deportation; and to say a proper good-bye to the Kalahari.

As we had done in 1985, Mark flew our plane from Gaborone and I drove to Deception Valley through the thorn country. In 1988 the drought that had gripped the desert since 1979 still had its hold on the land. The heavy clay soils of the ancient riverbed had been reduced to a sickly dry powder from the years of angry sun and tireless winds. As I drove across Deception and around Acacia Point, billows of pale dust rose behind the truck. Not a single gnarled scrub or tortured blade of grass had survived. Even the grass stubbles were long since buried under a layer of time.

Mark was standing by the plane, parked in the dilapidated boma we built years ago to keep the Blue Pride lions from chewing the tires. He walked out to greet me and we sat on the dry riverbed near our old camp and homestead. To the north was Acacia Point, Mid-Pan Water Hole, Eagle Island, Cheetah Hill, and North Dune; to the south, Bush Island, Tree Island, and Jackal Island in South Bay — all as familiar today as they were years ago. The Kalahari's face was ashen with drought, sandblasted, wrinkled, and lifeless from the harsh wind. I wondered if she had noticed a similar change in us; for we too had endured a long dry spell.

After a while we walked into what was left of our camp. The *Ziziphus mucronata* and *Acacia tortillas* trees had managed to produce leaves, but they were gray and withered. As I stepped into the barren tree island, a Marico flycatcher swooped past my head and landed on a branch five feet away. He chirped urgently, wings trembling at his sides as he begged for food. Surely this was one of the flycatchers we had known before. I ran to the cool box, tore off a piece of cheese, and held it out to him. He took it as if we had fed him only yesterday.

After we had been deported, friends had removed our tents, but the kitchen boma with its ragged thatch roof was still standing. The "hyena table" — built six feet off the ground to keep the brown hyenas from our pots and pans — remained beneath the acacia tree. The bath boma and a few stick tables lay in various stages of termite consumption. Tomorrow we would clean it all up, stack it into a huge pile, and burn it.

Late in the afternoon we dug up a bottle of Nederburg Cabernet Sauvignon that was still buried in our "wine cellar" beneath the ziziphus tree. We replaced it with a new bottle and a note in a jar, thinking that someday we might pass this way again. Then, sitting crosslegged on the riverbed, just a few yards from where our water drums once stood, we watched the sun disappear beyond the dunes. Sipping wine from tin mugs, we listened to the click-click-click of barking geckos and the mournful, wavering cries of a jackal somewhere beyond the dunes.

The next morning Mark was determined to search for the lions we had radio collared in 1985 on our initial return to the Kalahari, just weeks before we were deported. I didn't want to discourage him, but my own feeling was that there was very little chance of ever finding them. It had been two and a half years since we had darted and collared eight of them, including Happy, whom we had known for years, and Sunrise and Sage, her younger pridemates. Since our departure the lions must have scattered for thousands of square miles, searching for prey in the far reaches of the desert. But how could we not try to find them? We mounted the antennas under the wings, and Mark took off as he had hundreds of times, to search for lions in the dunes.

I stayed behind and began the grueling task of taking down the kitchen and bath bomas, the airplane fence, and the tables. I hacked away with an ax and dragged poles and grass to a large pit. The heat around me was intense, something you could almost reach out and touch. As I worked, I wondered how we had endured this for so many years. Sometimes I could hear the drone of the plane, as Mark flew on and on — north, south, east, and west on imaginary grid lines across the sky. When he landed for a lunch of nuts and fruit, his face was red with heat, grim and determined. All morning he had heard nothing through his earphones except static and phantom signals. He tried for hours again that afternoon, and landed just before sunset, shrugging his shoulders. Nothing.

The next morning he took off and flew until noon, landed to refuel, and took off again. I burned poles and cleared rubbish. I was beginning to think that hauling heavy logs through the heat while Mark waffled around in the cooler altitudes was a lousy arrangement. Then I heard the plane making a beeline for camp. It landed. Mark jumped out and ran toward me, shouting, "I've found them! You're not going to believe it. I've found seven of the eight lions! And Happy is just beyond East Dune."

As though we had never been away, we found ourselves once again in the truck, trundling over the dune faces, toward the beep-beep-beep of a lion's radio signal. This time my throat knotted, not from thirst but from the laboring of my heart. "Are you sure you saw her? Maybe she slipped her collar. Or maybe she's dead."

"Hang on, Boo. You'll see."

The dune grasses, bleached blond by the sun, rattled with drought as they swayed gently in the midday heat. The sun's rays beat straight down, making a desert without shadows. Mark stopped the truck and pointed ahead. Three lionesses sleeping under an acacia bush slowly raised their heads, panting as they peered at us. Two were very young, and we did not recognize them. But the other had a vaguely familiar face and wore a tattered radio collar. She stood, her eyes wide. It was Happy, now thirteen years old — ancient for a Kalahari lioness. We had sat with her for hundreds of hours, followed her across star-lit dunes, and even slept near her on the desert sands on several occasions. We had found Happy on our first return to the Kalahari in 1985, and here she was again.

She staggered to her feet and walked slowly toward us, her ribs jutting out in dark lines under skin like parchment, her belly high and tight to her sagging spine. She was old and she was starving. She stumbled, then hesitated, swaying, her strength dissipated in the waves of midday heat. She started forward again and came to within ten feet of the truck. I thought for a moment that she might chew the tires, as she and the other Blue Pride lionesses had done so often before. Instead, she looked at us with soft, golden eyes whose lack of fear told us all we needed to know about whether or not she recognized us. If only we could ask her, "Where's Blue? Whatever happened to Moffet? Are these your cubs?" But years had gone by, and the answers were forever lost in the desert.

She walked around the truck, nearly touching its rear bumper with her side, and rested again in the patchy shade of a few scrawny branches. We stayed with the lionesses all afternoon and that evening watched them try unsuccessfully to kill a large male gemsbok. The effort exhausted them and they stood panting, their shrunken bellies heaving. They were a pitiful trio: Happy too old to hunt well anymore, the young ones too inexperienced. Born in a land that only rarely offers water and is stingy even with shade, they stood like three spindly monuments to survival. We left them at sunset, feeling the way the Kalahari always makes us feel; intrigued by her wonders, sobered by her harshness, saddened by her finality.

The next morning we drove across East Dune and found Happy one final time. She was with three older females as well as the two younger ones: the Blue Pride members, unnamed, unknown, but enduring. Tucked under an *Acacia mellifera* bush was a freshly killed gemsbok. All of the lions' bellies, including Happy's, were full.

Animals know about greetings: long-separated lions rush to one another, rubbing heads and bodies together in reunion; brown hyenas smell one another's necks and tails; jackals sniff noses. But do animals know about good-bye, I wondered as we drove away from Happy for the last time. I held her eyes with mine until her tawny face faded into the straw-colored grasses of the Kalahari. We knew we would not see her again; she would never survive another dry season. But at least she was surrounded by her pride in a reserve that was secure.

We returned to North Luangwa to continue our programs there, and on landing at the airstrip we were told by the guys that a lioness had been visiting their camp every evening. Several nights later we found her near the strip, where we darted her. And as always we weighed, collared, and named her. She is small, not very strong, but she is still here. She is Hope.

Postscript

DELIA

WHEN WE BEGAN our project in 1986, the elephants of North Luangwa National Park were being poached at the rate of one thousand each year. By the end of 1991 that number had been reduced to twelve.

On January 16, 1992, David Chile of Mwamfushi Village, one of the scholarship students of our North Luangwa Conservation Project, presented newly elected President Frederick Chiluba with a petition signed by three thousand Zambians requesting that their government observe the ivory ban. Many other organizations and individuals in Zambia — the David Shepard Foundation and the Species Protection Division are but two examples — participated in the effort to convince the Zambian government to join the ban. On Save the Elephant Day, organized by Wanda and Tom Canon, students all over Zambia and in the United States sang a special song they had written about the elephants. The singing was coordinated so that it lasted for five hours across Africa and America. The song was heard.

In a press statement on February 7, 1992, President Chiluba's government announced that it would join and fully support the international ban on ivory trade: "After reviewing evidence of a disastrous decline of the country's elephant population under the previous government, [the new government] has announced a radical change in Zambia's elephant policy . . . Zambia effectively is opposed to the resumption of international trading in elephant products." The new minister of tourism went on to request that other African nations still trading in ivory follow Zambia's lead, and he invited their cooperation and coordination in establishing measures that would ensure the conservation of their collective wildlife resources. Finally: "On Friday, 14th February 1992, the

Minister has arranged for a ceremonial burning of ivory seized by National Parks and Wildlife Services and other Zambian agencies from poachers and smugglers."

o o o

The North Luangwa Conservation Project (NLCP) is now housed in a smart office in Mpika. Evans Mukuka, our current education officer, visits ten schools a month, presenting slide shows about wildlife and conservation to the children. These programs, which first began on the red, dusty clay of Chishala Village, now reach twelve thousand students in thirty schools, in villages that once were notorious for poaching. Our scholarship program sponsors a student from each village to attend the University of Zambia.

Recently Mukuka held a wildlife quiz competition among the sixth-graders of the Mpika area schools. Teams from each school answered a battery of questions about the wildlife of North Luangwa — and the children of Mwamfushi won!

With the assistance of the Canons, Max Saili, Ian Spincer, and Edward North, our village programs help people find new jobs, start cottage industries, and grow more protein. In all, the NLCP has created more than two hundred jobs for local men and women, many of whom were once involved with poaching.

In much of Mpika District, for many years there were no butcher shops or other places to buy domestic meat. So the Bemba people, who have a long history of subsistence hunting and a keen desire for meat, were poaching wild animals to extinction. As a way of discouraging subsistence and commercial meat poaching, the project loaned enough cash to a Zambian businessman in Mpika to open a butcher shop. He brings cattle up by train from the Southern Province, butchers it, and sells beef at a lower price than that charged by black marketers who sell poached meat. A sign on the side of his butchery urges, SAVE WILDLIFE: BUY BEEF, NOT BUSH MEAT.

Confronted by the American ambassador, the official who had swallowed Bwalya Muchisa's story and charged Mark with buying black-market military rifles backed down. Bwalya has disappeared.

The Anticorruption Commission has formed a Species Protection Division (SPD), which is charged with looking into official corruption related to poaching. Periodically, SPD officers come to Mpika to investigate officials who collaborate with poachers. In 1991 they arrested Mpika's chief of police, the police station commander and an armory officer at Tazara, and two tribal chiefs. In early 1992 Warden Mulenga was discharged and Isaac Longwe, a very capable and trustworthy man, was made acting warden.

In spite of the progress, we cannot yet claim that North Luangwa is secure. Corruption is still unbridled, although under President Chiluba's administration we have renewed hope that it will diminish. Poaching continues, though it is much reduced. The police released Simu Chimba, the "little" big poacher captured by the Mano scouts; but later he was killed in the Zambezi Valley — by a charging elephant. Chikilinti eludes the scouts with his powers of invisible juju; these days, however, he poaches more often in the game management areas outside the park.

Pressured by former warden Mulenga to join his corrupt activities, Kotela — who so transformed the Mano Unit — requested a transfer to another post. We will be sad to lose him, as ultimately the protection and development of the park depend on him and his countrymen. It is up to them, not us, to make it work.

In October 1991 the new democratic government in Zambia, the MMD (Movement for Multiparty Democracy), was voted into office by an overwhelming majority — in a free and fair election monitored by former U.S. President Jimmy Carter. President Chiluba's administration is committed to the conservation of Zambia's natural resources, including its wildlife, and is supportive of a free-market system that welcomes visitors to Zambia. Many of the high-level officials of the previous one-party system who were heavily involved in poaching have been voted out of office, and the new government is addressing the widespread problems of corruption and exploitation. Poaching is so institutionalized that it will take some time before the administration can stamp it out. Nevertheless, Zambia has an opportunity to start over, and to demonstrate that man and wildlife can live side by side for the benefit of both.

The country will need strong support in this quest for a truly effective national conservation program.

Until substantial benefits can be realized from tourism and other wildlife-related industries, the North Luangwa Conservation Project must continue to find ways of fostering an economic bond between the park's animal communities and nearby villages, which might otherwise destroy the wildlife. Unfortunately, if our project and its community services were to disappear tomorrow, poaching would again threaten the park. The short-term advantages to the villagers eventually must be replaced by sustainable benefits that come directly from the park.

Tourism may be the answer, but it must be designed so that it does not disrupt the ecosystem. We have strongly recommended that it be limited to old-fashioned walking safaris. Everything is ready: the park has been secured from commercial poachers, the tour camps are set up, and there will soon be an official way to return money from tourism to the Bemba and Bisa people. All that remains is for tourists to come.

They have started. Two small companies have established walking safaris in the park. Their visitors do not have to ride in radio-controlled minibuses and elbow their way through crowds to see lions on a kill. Each person who comes and walks in the real Africa helps save elephants by making living wild animals valuable to the local people.

o o o

In the meantime Mark and I will continue in North Luangwa, assisting the government in its responsibility to secure, manage, and develop the park for the benefit of people and wildlife.

Simbeye, Mwamba, and Kasokola, who joined us at the very beginning, are still with us and still smile every morning. They and the other members of our Zambian team are working with spirit and determination to save North Luangwa.

Bouncer and his pride continue to move from the plains in the wet season to the forest near Nyama Zamara lagoon in the dry season. The Serendipity Pride maintains its territory along the Mwaleshi and tries to avoid a certain crocodile. Mona, the monitor lizard, has abandoned our bathtub and made a new nest in the side of the riverbank nearby.

Sometimes Survivor and Cheers come by our camp as they migrate to and from the mountains. On rare occasions Survivor ventures into our camp at night to feed on the marula fruits. He walks as softly as before, slurps the fruits as loudly as ever, and lulls us to sleep with his song.

Appendixes

Notes

Acknowledgments

Index

Fences and Kalahari Wildlife

o o o

In the Kalahari Desert, fences are blocking antelope migrations and extinguishing wildlife populations. The fences are being erected (a) to control foot-and-mouth disease (FMD) and (b) to enclose large commercial cattle ranches.

FOOT-AND-MOUTH DISEASE QUARANTINE

In some cases fences constructed for this purpose run for hundreds of miles across the savanna. They were built along the southern, western, northern, and part of the eastern boundary of the Central Kalahari Game Reserve, blocking antelope migrations to and from the Botetle River and Lake Xau, the only natural watering points for animals during drought.

These fences divide Botswana into quarantine sections, so that in the event of an FMD outbreak diseased cattle can be isolated, theoretically preventing spread of the infection to another sector. Another purpose of the fences is to segregate wildlife populations suspected of harboring the disease from domestic stock.

After years of research in the Kalahari we questioned the efficacy of these fences for several reasons:

1. No FMD virus has ever been found in Kalahari wildlife.[1]
2. During FMD outbreaks the virus often spreads from one quarantine area to another, irrespective of the fences — which therefore do not seem to be effective barriers. Furthermore, in Europe FMD virus has been carried in damp soil on the feet of birds and rodents, on the wheels of vehicles, or even in the air.[2] Posts and wire cannot contain the movement of such vectors, and so the disease spreads across the fences.
3. It has never been proved that wild ungulates can transmit FMD to domestic stock.[3]

In spite of the fact that the fences do not control FMD, since they were erected in the early 1950s nine major desert antelope die-offs have occurred, similar to the one we witnessed near Lake Xau. In at least five consecutive years, beginning in 1979, massive extinctions of migratory wildebeests were recorded at Lake Xau (the drought and the DeBeers diamond mine had pumped the lake dry) by the two of us and by Doug Williamson, who manned our Deception Valley camp from 1981 to 1984. Rick Lamba, a film producer, also witnessed the tragedy and made a documentary about it entitled "Frightened Wilderness." The film was aired on the Turner networks and shown on Capitol Hill. The numbers of animals perishing at Xau each year varied from fifteen thousand to sixty thousand, but eventually more than a quarter of a million wildebeests died.

But the dying wildebeests at Lake Xau were only part of the story. Sixty miles south, up to ten thousand red hartebeests were dying each year, along with uncounted numbers of gemsbok, giraffes, springbok, and other desert antelope. They piled up and died against fences that kept them from water. In total since the 1950s, the fences have killed more than a million wild ungulates, and additional numbers of carnivores that depend on them as prey.

The fences, although merely the front line of exploitation, provided hard evidence that cattle-development money was decimating wildlife populations. Tractors pulling wagons loaded with armed men regularly patrolled the fences; any wild animal that came near was shot. Thousands did, and they were killed. One of the original owners of Safari South (Pty), a safari hunting company based in Maun, has a photograph of a pile of antelope bones "as large as a two-story house" taken near the northern end of the Makalamabedi fence line. According to a range ecologist for Botswana's Department of National Parks, these patrol crews made a business of marketing in Gaborone — the capital — meat and skins from animals killed along the fences. He reported seeing a mass grave full of hundreds of fresh carcasses on one of the government's experimental farms.

By any standard, these man-induced mortalities represent one of the worst wildlife disasters of this century — one that could have been avoided entirely.

ENCLOSURE OF LARGE COMMERCIAL RANCHES

Beginning in the 1970s, wealthy private cattle ranchers, including some of the major political figures in Botswana, were given low-interest loans from the World Bank to develop huge ranches in Kalahari wilderness

areas. They built fences, drilled wells, raised cattle — and blocked antelope migrations, killing tens of thousands.

Typically these ranches were profitable for about five years. Then the
wells, which were drawing water from fossil (unrechargeable) aquifers,
dried up; the semiarid savannas were overgrazed to scrub and dust.
Before the ranchers could repay the loans, they were forced to abandon
their ranches and move on to "develop" the next enormous plot of
wilderness — with another loan from the World Bank. This high-finance
version of slash-and-burn agriculture left in its wake sterile wastelands
covered with coils of fence wire and piles of bleached skeletons, the
remains of tens of thousands of antelope whose migrations to water had
been blocked.

It was only a matter of time before these commercial ranchers began
to run short of land. When they did, they proposed dissolving the Central
Kalahari Game Reserve, the second or third largest wildlife protectorate
on earth, so that they could use it for additional cattle ranches.

After destroying tens of thousands of square miles of wilderness habitat in the name of development, the ranchers never repaid any of the
World Bank's loans. Even so, the Bank was about to finance another
major cattle-development scheme in Botswana when an international
outcry stopped it. The United States, which provides about 20 percent
of the Bank's budget, balked at funding any more such projects in equa-
torial Africa without proper environmental controls.

Botswana's commercial cattle industry has been profitable in the short
term only because the European Community (EC) countries have paid
60 percent above the world price for Botswana beef and have guaranteed
to import as much as the ranchers could produce. But only 3 percent of
Botswana's households are getting two thirds of the profits from the
industry. Meanwhile, the Common Market countries were paying exor-
bitant sums to keep frozen a 720,000-metric-ton surplus of beef. Eventually, to reduce the surplus, much of the beef would be sold to Russia
for 10 percent of the cost to produce it. Furthermore, the EC rebated to
Botswana's commercial ranchers 91 percent of the tariff charged for
access to its market. Lavish low-interest development loans by the World
Bank, coupled with high returns from the EC, created a powerful incen-
tive for ranchers-cum-politicians in Botswana. They cashed in by developing huge ranching blocks in wilderness areas, regardless of the cost to
the environment.

On our return to the Kalahari in 1987 we flew over the northeast
quarter of the reserve. There we discovered hundreds of cattle and goats

grazing up to twenty miles inside the reserve, where they were being watered by game scouts — at boreholes developed with E.C money for migratory wildlife. The spread of cattle into the reserve had begun, which may be the reason we were ordered to leave the Kalahari.

In the end, however, as described in the text, the government of Botswana voted to keep the reserve intact and not develop it for cattle.

The Ivory Ban

o o o

POACHING OF THE AFRICAN ELEPHANT BEFORE 1989

From 1963 to 1989 poachers shot 86 percent of the elephants in Africa for their ivory, skin, tails, and feet. In one decade the population plummeted from 1,300,000 to 600,000 — less than half its former size. Seventy thousand elephants were shot every year to meet the world's demand for ivory. Ninety percent of the ivory entering the international market was from poached elephants. In other words, there was a 90 percent chance that an ivory bracelet in any jewelry or department store in the world was from a poached elephant. Illegal tusks were "laundered" by using false documents.

The elephant populations in twenty-one African nations declined significantly in the decade preceding the CITES ban. Zambia lost more than 80 percent of its elephants. In the Luangwa Valley alone, 100,000 elephants were shot in the period between 1973 and 1985. In North Luangwa National Park from 1975 to 1986, elephants were shot at the rate of 1000 per year. Tanzania lost 80 percent, Uganda 73 percent. In East Africa as a whole, 80 to 86 percent of the elephants were shot by poachers. In the fifteen years prior to the ban, Kenya lost 5000 elephants a year — 1095 a year in Tsavo Park alone.[1]

In 1986 the United Nations Convention for International Trade in Endangered Species (CITES) attempted to control the illegal ivory trade by requiring that import-export documents written in indelible ink accompany each tusk. This scheme failed completely. In some instances documents and marks were falsified; most of the time the procedure was ignored altogether. The poaching of elephants and the illegal ivory trade continued as before.

Most African nations did not have the resources to control elephant poaching. The price of raw ivory soared to more than $136 per pound. Corrupt government officials in many nations (including Zambia, Zimbabwe, Botswana, and South Africa)[2] participated in poaching to supplement their incomes. Governmental institutions — customs, the army, police, departments of National Parks, the judicial system — frequently were involved as well. In the majority of cases the real profit from poached elephants went into the hands of private individuals or to foreign nationals. The national treasuries of the African countries benefited very little.[3]

THE CITES BAN

In early 1989, realizing that poaching was out of control and that their nations were losing a tremendous resource, eight African countries (Tanzania, Kenya, Somalia, Gambia, Zaire, Chad, Niger, and Zambia) agreed to support an international ban on the ivory trade, to begin in January 1990. In March 1989 the United States imposed an immediate ban on the importation of ivory. Canada, the EC, Switzerland, and the United Arab Emirates followed the U.S. lead.

On October 17, 1989, CITES voted (seventy-six nations to eleven) to list the African elephant on Appendix I, thereby declaring it an endangered species. The sale of all elephant parts was prohibited for two years as of January 1990.

The price of ivory paid to poachers plummeted to a hundredth of its former price: about $1.36 per pound. As a result, poaching decreased dramatically in many areas: Kenya, which had lost 5000 elephants a year, reported only 55 shot the year following the ban. In North Luangwa National Park we recorded only 12 dead elephants in 1990 (down from 1000 a year). In Selous, Tanzania, no fresh carcasses were observed. In general, poaching declined in East Africa by 80 percent. It also decreased in Chad, Gabon, Zaire, and Congo.[4]

These incredible results show that the CITES ivory ban was one of the most effective environmental policies ever adopted.

RESISTANCE TO THE BAN

In spite of this unprecedented success, eight nations who stood to gain financially from the ivory trade filed reservations to the ban in 1989: China, Botswana, Zimbabwe, Mozambique, South Africa, Malawi, Zambia — which had changed its position — and Great Britain on behalf of Hong Kong (for six months).

Furthermore, South Africa, Botswana, Zimbabwe, Namibia, Malawi, Mozambique, and the previous government of Zambia moved to downlist the elephant from "endangered" status to "threatened" and to continue their ivory trade. With the exception of South Africa, these nations formed their own ivory cartel, the Southern African Center for Ivory Marketing treaty (SACIM).

In an extremely courageous move, the newly elected government of Zambia, under the MMD party, announced that it would change its position again and support the ban, and not sell the stockpile of ivory it had confiscated from poachers. On February 14, 1992, it staged a ceremonial burning of the illegal ivory. Also, China reversed its position and joined the ban.

The Southern African nations that refused to join the ban mounted an international campaign to convince the non-African nations of CITES that their position of down-listing the elephant was the correct one. Their arguments were flawed for the following reasons.

1. These nations are involved in the illegal ivory trade.

South Africa, one of the most outspoken of the nations resisting the ban, is one of the largest clearinghouses for illicit ivory on the African continent. Raw ivory entering and leaving South Africa for other countries in its customs union (Botswana, Swaziland, Namibia, Lesotho) does not require import or export permits. In addition, worked ivory can be imported to or exported from South Africa without permits. So the door is wide open for illegal ivory to be imported into the country, then exported anywhere in the world without documents.

Much of the illicit ivory from Zambia, including that from North Luangwa National Park, has been flown on Swazi-Air from Lusaka to Swaziland, then trucked to South Africa. Three top-level Zambian officials were suspended for participating in this smuggling ring.

Once the ivory is in South Africa, it can be freely sold or exported. Before the 1989 ban South Africa imported 15 tons of illegal ivory from Zaire, 12 tons from Angola, 10 tons from Zambia, 2 tons from Zimbabwe, and 1 ton each from Malawi and Mozambique.[5] And South Africa exported 40 tons of illegal ivory annually. A United Nations study concluded that "South Africa serves as a conduit for the illegal export of significant quantities of ivory . . . from neighboring states (including Angola, Botswana, Malawi, Mozambique, Zambia, and Zimbabwe)."[6]

In Angola, members of the National Union for the Total Independence of Angola (UNITA) killed 100,000 elephants to finance their war with the government. These tusks were exported to South Africa, where they entered the free market.[7] A photographer from *Time* magazine witnessed a huge machine shop run by UNITA in Angola, where dozens of lathes were being used to carve tusks into replicas of machine guns (personal communications). Two men arrested in South Africa in possession of 975 poached elephant tusks were not prosecuted.[8]

On February 25, 1992, the Environmental Investigation Agency of England reported that after two years of scrutiny it had determined that the South African Defense Force (SADF) and the Zimbabwe National Army had been involved in large-scale elephant poaching and ivory smuggling. Their report provided evidence that the SADF ran a major smuggling operation out of Angola and Mozambique.[9]

2. Claims that there are too many elephants in some areas are inaccurate or irrelevant.

Zimbabwe and Botswana declare that they have too many elephants and need to cull them to prevent habitat destruction. Too often when elephants appear to occur in high numbers, it is actually because they have been crowded into small areas by outside poaching pressures, or by loss of habitat from human development. If the poaching were eliminated or if elephants were allowed to inhabit a greater portion of their former ranges, they would no longer be overcrowded. With human populations growing more than 3 percent annually, people will take over more and more elephant habitat for development and conflicts will occur. But the CITES treaty does *not* prohibit the culling of elephants in areas where their densities are too high. Culling should be considered a last resort, but when necessary it can be done according to the CITES regulations. Culling does not cause poaching; selling the ivory and other parts from culled elephants does.

Too often in the past, governments have repeatedly and prematurely resorted to culling operations to control elephant densities. It would be far more appropriate for the central, southern, and east African nations to form an international policing agency similar to Interpol to deal with the illicit traffic in animal parts and to coordinate antipoaching law-enforcement operations.

3. These countries claim that they have controlled poaching and that by being denied a trade in elephants' parts they are being penalized for the lack of control in other nations. But poaching still continues in Zimbabwe and Botswana. According to the deputy director of Zimbabwe's Department of Wildlife, the number of elephants killed by poachers has increased 300 percent in recent years.[10] The department reported intensive organized elephant poaching in the Zambezi Valley and was granted

$104,500 from the United States to help control it. Elephant poaching occurs in Chobe Game Reserve in Botswana and Nigel Hunter, the deputy director of Wildlife and National Parks there, has stated that "he suspects that illegally-taken ivory from Botswana moves through South Africa." [11]

Because there is at present no way to verify the origin of elephant parts, any country that trades in them inevitably stimulates a massive illegal traffic within its own borders and across its borders with other countries. As the demand for illegal parts grows, more and more elephants are poached in neighboring countries to fill the market.

4. Claims that people should be able to benefit from elephants are not contradictory to the ban.

The countries resisting the ban state that local people and national treasuries should be able to benefit from elephants. But the ban does not prevent them from doing so. Wildlife tourism can generate as much income as the ivory trade. Kenya's living elephants bring in $20 million a year through tourism, which flows to many different people; money from poached elephants falls into only a few hands.

5. Some nations want to down-list the African elephant so that there can be an international trade in hides, feet, and tails, even if trade in ivory is prohibited.

Before the ban, the trade of elephant skins, feet, and tails was worth as much as the ivory trade in Zimbabwe and in South Africa. It matters little to the poacher whether he shoots an elephant for its ivory or its skin. As long as there is *any* market in *any* elephant parts, poaching will increase again.

BAN RECONSIDERED

In March 1992 the CITES delegates met in Japan to vote on whether or not to continue the ban. In spite of its success, the SACIM nations and South Africa wanted it reversed. Unbelievably, the delegation from the U.S. Fish and Wildlife Service considered joining these countries in their vote for a down-listing of the elephant. Apparently the U.S. delegates had accepted statements by the SACIM nations that they had the means to control such a trade, despite overwhelming evidence that they could not. Sixty American environmental and conservation groups, including our own, as well as many senators and congressmen, sent petitions to

President George Bush, asking that the United States support a continued moratorium. After weighing all the evidence, President Bush decided that this action was "the right thing to do." He instructed the U.S. delegation in Japan to vote accordingly, and other CITES member nations followed suit. In the end, the nations opposing the ban withdrew their proposal and the moratorium on the sale of all elephant parts was continued.

CONCLUSION

Opening a legal market for *any* elephant parts (ivory, skins, tails, feet) will reopen the illegal market. Under the present conditions of widespread corruption and lack of resources to protect elephants in the field, *the most effective way to save the African elephant is with a continued long-term, complete international moratorium on the sale of all elephant parts, including ivory.* The next time CITES meets (in 1994), the moratorium should be extended for at least ten years, not two. A longer period will prevent stockpiling of ivory by poachers and send a strong message to black-market dealers that the CITES nations are committed to saving the African elephant from extinction. It will also permit elephant populations to rebuild their numbers.

One final word: the ivory trade not only kills elephants but also leads to the deaths of people trying to protect them.

APPENDIX C

Large Mammals of North Luangwa National Park

o o o

Baboon, chacma	*Papio ursinus jubilaeus*
Bush pig	*Potamochoerus porcus*
Bushbuck	*Tragelaphus scriptus*
Cape buffalo	*Syncerus caffer*
Cheetah	*Acinonyx jubatus*
Duiker, common	*Sylvicapra grimmia*
Eland	*Taurotragus oryx*
Elephant	*Loxodonta africana*
Giraffe, Thornicroft's	*Giraffa camelopardalis thornicrofti*
Hartebeest, Lichtenstein's	*Alcelaphus lichtensteini*
Hippopotamus	*Hippopotamus amphibius*
Hyena, spotted	*Crocuta crocuta*
Impala	*Aepyceros melampus*
Kudu	*Tragelaphus strepsiceros*
Leopard	*Panthera pardus*
Lion	*Panthera leo*
Monkey, samonga	*Cercopithecus alboqularis*
Monkey, vervet	*Cercopithecus pygerythrus*
Oribi	*Ourebia ourebi*
Puku	*Kobus vardonii*
Reedbuck	*Redunca arundinum*
Roan antelope	*Hippotragus equinus*
Sable antelope	*Hippotragus niger*
Warthog	*Phacochoerus aethiopicus*
Waterbuck	*Kobus ellipsiprymnus*
Wild dog	*Lycaon pictus*
Wildebeest, Cookson	*Connochaetes taurinus cooksoni*
Zebra	*Equus burchelli*

Notes

CHAPTER 9

1. Cynthia Moss, *Elephant Memories* (New York: Random House, Fawcett Columbine, 1988).
2. Ibid.
3. Ibid.

APPENDIX A

1. R. S. Hedger, Foot and mouth disease, in *Infectious Diseases of Wild Mammals*, ed. John Davis et al. (Ames: Iowa State University Press, 1981).
2. Ibid.
3. J. B. Condy and R. S. Hedger, The survival of foot and mouth disease virus in African buffalo with nontransference of infection to domestic cattle, *Research in Veterinary Science* 39(3):181–84.

APPENDIX B

1. Comments of the Humane Society of the United States Regarding Proposals by Zimbabwe, Botswana, South Africa, Namibia, Malawi, and Zambia to Transfer Populations of the African Elephant from CITES Appendix I to II (Humane Society of the United States, January 30, 1992).
2. Ibid.
3. African Wildlife Foundation, personal communication.
4. Comments of the Humane Society.
5. *The Ivory Trade in Southern Africa*, CITES Document 7.22, Annex 2, 1990.
6. United Nations Environmental *Panel of Experts Report*, August 16, 1991.
7. Craig van Note, *Earth Island Journal*, 1988.

8. *Johannesburg Star,* November 19, 1989.
9. *Under Fire: Elephants in the Front Line* (London: *Environmental Investigation Agency, 1992*).
10. *New African,* June 1991.
11. Comments of the Humane Society.

Acknowledgments

We are extremely grateful for the conservation initiatives taken by Zambia's new government, and especially for the vision of Frederick Chiluba, its new, democratically elected president. For the first time, we dare to hope that solutions to wildlife and human development problems may be complementary and lasting. Our thanks to the government of Zambia and the Mpika District Council for allowing us to conduct the North Luangwa Conservation Project; to former U.S. ambassador to Zambia Paul Hare and U.S. Information Services officer Jan Zehner for their roles in securing this permission; and to their wives for opening their homes to us on many occasions. We are indebted to the current ambassador, Gordon Streeb, for adding his prestige and influence to the politics of conservation in Zambia, and we appreciate the hospitality he and Junie, his wife, have shown us.

To Andy and Caroline Anderson and Dick Houston in Lusaka, thanks for understanding what it means to come in from the bush. Also for their friendship and hospitality in Lusaka, we thank Julie and Alan van Edgmond, Mary Ann Epily, Marilyn Santin, and Mary, Ralph, and Astrid Krag-Olsen.

We owe special thanks to Luke Daka, permanent secretary to the minister of tourism; to Akim Mwenya, director of National Parks and Wildlife Services of Zambia; to Gilson Kaweche, also of NPWS; to Paul Russell, operations head of the Anticorruption Commission; and to Norbert Mumba, Clement Mwale, and Charles Lengalenga, chief investigative officers of the Species Protection Division for their help in breaking up organized poaching in the Mpika area. Our sincere appreciation to the government of Canada for donating a truck to our community service program.

Our heartfelt thanks to President George Bush, and First Lady Barbara Bush, for their time and consideration in listening to our message and acting decisively to ensure continuation of the CITES international ban on trading elephant parts. We are grateful also to Senator Bob Kasten, to Eva, his wife, and to Alex Echols, the senator's senior staffer; and to Congressman Mel Levine and staffer Jennifer Savage for their actions to guarantee that subsidies from the World Bank do not continue to destroy Kalahari wilderness, and again to ensure that the U.S. delegation to CITES voted to continue the ivory ban. Our sincere appreciation also to Marguerite Williams for her assistance in this regard, and for all the other ways she has helped us.

We are very grateful to the Friends of the Animals of the Frankfurt Zoological Society, and especially to Richard Faust, its president, and Ingrid Koberstein, his assistant, who since 1978 have been the major supporters of our projects and have provided everything from paperclips to airplanes. Frankfurt is a strong force for conservation in Africa, Asia, and South America.

We are equally indebted to the members of our own Owens Foundation for Wildlife Conservation, who have supported us financially, morally, and spiritually, especially in the expansion of our North Luangwa Conservation Project's law enforcement, village outreach/community service, and conservation education programs.

To Helen Cooper, Delia's sister and our foundation's executive director, and to Fred, her husband — thanks for being there with everything from direct contact with presidents to deft leadership, sage advice, fund raising, and the organization of our speaking tours. Nephew Jay Cooper is our computer consultant and all-around hack; thanks, Jaybird. Thanks also to Jay's brother, Derick, for letting us store our lion range plots on his bed and on the walls of his bedroom. Our warmest thanks to Bobby Dykes, Delia's twin brother, for managing our photographic library. Mary, his wife, is the assistant director of our foundation and acquires items as diverse as airplane engines and crayons. One way or another, she gets them to one of the most remote corners of Africa; even more miraculously, she accounts for their purchase and shipment in as many as five different currencies. At the same time, she heads our program of sister schools. Additional thanks go to Mark's brother, Mike, and Jan, his wife, for their care of photographic materials.

Leslie Keller-Howington donated the beautiful artwork for our foundation's logo and lecture brochures. We are also very grateful to Rick Richey for video editing and reproduction and to Channing Huser for her illustrations for the North Luangwa Conservation Project.

For months volunteer Marie Hill worked in Mpika and more remote villages to develop our conservation education program. Harvey Hill provided fresh ideas and material support. We are indebted to all of them. When our programs were fast growing into a regional development project, Tom and Wanda Canon, our volunteer project coordinators in Mpika, came along to add order, calm reassurance, and confidence to the entire effort. Even though the sun dims and the thatch on their roundhouse smolders whenever they switch on their mega-appliances, they have helped the project tremendously and we could not do without them. To Max Saili, our fine community service officer, and Evans Mukuka, our education officer, we extend deep gratitude for helping take the message of conservation to the people of Mpika District.

Ian Spincer and Edward North, who joined the project as young graduates from the University of Reading in England, have literally waded flood-swollen, croc-infested rivers to get reliable firsthand information on poaching and game-guard field performance. No risk has been too daunting, no challenge too great, no task too menial for them to tackle. And through it all they have helped restore our sense of humor.

Our thanks also to Christopher, Mark's son, who for three months in 1991 helped build tracks, drove trucks, and named Bouncer, the lion.

Very special thanks to Mick Slater, David and Jane Warwick, Dutch Gibson, Barbara Collinson, and to Glen Allison, Charlotte Harmon, Gracious Siyanga, Grace, Exilda Mungulbe, Patrick Enyus, Carl Berryman and all the staff of the District Development Services Project (Masdar; British Overseas Development Agency) in Mpika who have sheltered us in their homes, allowed us to use their communications equipment and post office space, and encouraged and supported us in ways too numerous to mention.

We are grateful to officers Isaac Longwe, Martin Mwanza, and Mukendwa Kotela for bringing diligence and integrity to the Mano game guard unit.

Steve Hall, of Wings of Eagles, Tampa, Florida, a renowned ferry pilot, flew our plane all the way from Atlanta to Marula-Puku camp. He often flies to Africa, and whenever he comes within a few hundred miles of Zambia, Steve soars in, his plane engorged with supplies for the North Luangwa Conservation Project. To a great guy in the sky, thanks, Steve — and fly safely.

Students, teachers, and parents from more than thirty American schools support conservation in Zambia by sending letters, art work, stories, and reports and by donating books and school supplies to their friends in the sister schools.

We applaud Judith Hawke, our Lusaka coordinator, for her efficiency and diligence in expediting the flow of permits and other bureaucratic paperwork as well as personnel, information, equipment, and supplies for the project.

We are grateful for donations of video equipment from the Sony Corporation, and from Bubba, the Zenith dealer in Portland; for computer equipment from Hewlett Packard South Africa and Kaypro of California; and especially to our close friend Jose Jardim for years of computer support and for helping us clean up our camp in Deception Valley. We are also indebted to Richard Ferris of Kodak South Africa for stocks of film, and for hospitality and friendship.

For their helpful comments on the manuscript we thank Bob Ivey, Dick Houston, Lee and Maureen Ewell, Jon Fisher, Barbara Frybarger, Barbara Brookes, and Helen Cooper. To dear friends Bob Ivey and Jill Bowman, thanks once again for allowing us to rattle around your home spoiling your cats while putting the finishing touches to the manuscript.

Special thanks to Harry Foster, our friend and editor at Houghton Mifflin, for his enthusiasm, encouragement, and tireless editing of countless drafts of the manuscript; to Vivian Wheeler for her assistance; and to Suzanne Gluck, our literary agent, for her interest and support.

Dave Erskine and Gordon Bennet of Johannesburg evacuated our camp after we were deported from Botswana. In Johannesburg, George and Penny Poole, and Nick and Sally, kindly allowed us to stay in their family's A-frame cottage for three months after our deportation, and to recuperate at their cottage on the South Coast. Our old friend Kevin Gill offered us his home, companionship, and invaluable assistance in identifying the trees of North Luangwa. Everard and Patsy Reed of Johannesburg invited us to share their beautiful farm near Mulders Drift during our initial writing of the manuscript.

Hank and Margaret McCamish, we thank you for that very special place in the valley of the deer, and for your faith and trust in us and our philosophy of helping the people and animals of wild Africa.

We feel a special sense of comradeship with our "guys" at Marula-Puku: Mutale Kasokola, Mumanga Kasokola, Chanda Mwamba, Chomba Simbeye, Evans Mulenga, Timothy Nsingo, and other members of the A-team who have labored hard and risked much to save North Luangwa. Without them the park surely would have been lost by now.

To Bill Campbell and Maryanne Vollers, thanks for all the memories under the African stars and for your never-ending efforts for conservation. Bill and Marion Hamilton, Joel Berger, and Carol Cunningham were always there when we needed them. We are grateful to Jim and

John Lipscomb, producers of "African Odyssey," for following us around the continent on a seemingly endless journey of abandoned campsites, which ended with the beginning of a new dream.

Thank you to Randy Jones and Jim Cole for creating our brochure and to Joe and Geri Naylor for their help in producing it.

To all of our friends and family mentioned above, and to any we may have omitted, thanks for being a special part of our lives — and for helping to save the elephants of North Luangwa.

Index

Acacia trees, 90; *albeda*, 81; *mellifera*, 269; *tortillas*, 6, 26, 266
Acacia Point (Botswana), 14, 24, 266
Adamson, George, 183–84
Agricultural projects, 32–33, 185. *See also* Cattle-raising
Air patrols, 85–86, 124, 144–52, 201–8, 215, 220, 256; effectiveness of, 154, 167–72, 173–75, 195–96, 242, 257; poacher camps blasted, 235–38, 239–40, 241, 247
Airstrip One (Zambia), 53, 60–69 *passim*, 75
ANC (African National Congress), bombing, 34, 35, 36
Angola, 33, 39; ivory trade, 198, 285, 286
Antelope, 15; die-off of, 14, 16, 31, 279–80, 281; eland, 45, 57, 102, 212; hartebeest, 18, 45, 280; kudu, 103, 177, 212; roan, 45; sable, 45, 97, 100; steenbok, 25; *See also* Puku; Waterbuck
Anticorruption Commission. *See* Zambia
Arius (Zambian tribesman), 46

Baboons, 60, 129, 212
Badgers, 61
Banda, Mr. (officer from Zambia Division of Civil Aviation), 74–75
Bell, Alexander Graham, quoted, 32
Bemba tribe, 47, 74, 112, 113, 120, 123, 165; benefits to, and conservation, 61, 272, 274 (*see also* Conservation projects); as camp helpers, 70, 78, 84, 90, 109, 210, 211, 212; Chibemba language, 85, 86, 92, 107, 111–19 *passim*, 147, 197, 251; origin of tribe, 68

Berghoffer, "Bergie," 12, 15
Birds, 15–16, 45, 81, 82–83, 85, 177; at waterholes, 103, 213; *See also individual varieties*
Bisa tribe and language, 51, 61, 112, 113, 165, 274
Black market, 158, 228, 229, 243, 272
"Blue" (lion), 4, 25
Blue Lagoon National Park (Zambia), 39
Blue Pride. *See* Lions
"Bones" (lion), 4
Botetle River, 9, 279
Botswana, 3, 6, 9, 39, 265; air attacks on, 34, 35, 36; cattle-raising in, 31–33, 280–82; deportation of authors by, 28–30, 31–33, 34, 282; deportation reversed, 37–38, 266; fences in, 279–80; Immigration Department, 27, 28–30; ivory trade of, 179, 198, 283–87 *passim*; National Parks Department, 280, 287; *See also* Kalahari Desert
Botswana Development Corporation, 33
"Bouncer" (lion), 106, 206, 216–17, 274
Brachystegia. See Miombo trees
Britain: colonial government, 111, 112; and ivory trade, 284, 286
Buffalo, 59, 65, 79, 123, 132, 182, 212; Cape, 221, 222; confrontations with, 54–56, 96, 97–98, 225–26; lions prey on, 217; poaching of, 84, 161, 174, 250, 252, 253; at water holes, 45, 84, 103, 213–14
Bush, George, 32, 288
Bushbuck, 79, 182, 183, 214
Bushmen, 4, 8, 33
Bwalya (poacher and informant). *See* Muchisa, Bwalya

The Owens Foundation for Wildlife Conservation is a charitable organization which currently sponsors Delia and Mark's North Luangwa Conservation Project in Zambia. The project is attempting to recover North Luangwa Park and the surrounding wilderness areas from commercial poachers by strengthening game guard law enforcement programs, by sponsoring conservation education, and by involving local villagers in benefits from the area's wildlife. The project plans to expand its model programs to protect other endangered habitats and animals in Africa. If you would like to help in this important work, you may make tax-deductible donations to:

> The Owens Foundation for Wildlife Conservation
> P.O. Box 53396
> Atlanta, Georgia 30355

o o o

The authors are grateful for permission to quote from the following material: From "When We Looked Back" by William Stafford. Reprinted by permission. Copyright © 1959, 1987 The New Yorker Magazine, Inc. "Bivouac," from *Tickets for a Prayer Wheel*. Copyright © 1974 by Annie Dillard. Used by permission. "Heat," from *The Time Traveler*, by Joyce Carol Oates. Copyright © 1983, 1984, 1985, 1986, 1987, 1988, 1989 by The Ontario Review, Inc. Used by permission of the publisher, Dutton, an imprint of New American Library, a division of Penguin Books USA Inc. From "The Thrifty Elephant" by John Holmes. Reprinted by permission of Doris Holmes Eyges; © 1961, 1989 Doris Holmes Eyges. "The Eagle and the Hawk," by John Denver and Mike Taylor. Copyright © 1971 Cherry Lane Music Publishing Co., Inc. All rights reserved. Used by permission. "Sacagawea, Bird Woman" and "Something fragile, broken," from *Skin and Bones*, by Paula Gunn Allen. Copyright © 1988 by Paula Gunn Allen. Courtesy of West End Press. "Winter's End," from *A Season of Loss*, by Jim Barnes. Copyright © 1985 Purdue Research Foundation, West Lafayette, Indiana 47907. "Today I Am Docile" by Seth Richardson, from *Out of the Rain* by The Homeless of San Francisco. Copyright © 1988 by the St. Vincent de Paul Society. *Sacred Elephant* by Heathcote Williams. Copyright © 1989 by Heathcote Williams. Reprinted by permission of Harmony Books, a division of Crown Publishers, Inc. "Sweeney to Mrs. Porter in the Spring" by L. E. Sissman, in *Hello Darkness: Collected Poems* by L. E. Sissman, published by Little, Brown and Company. Poem by Walter Rinder from *Aura of Love*. Copyright © 1978 by Celestial Arts, Berkeley, California.

CPSIA information can be obtained
at www.ICGtesting.com
Printed in the USA
LVHW102133120122
708465LV00025B/702